ASTRONOMY'S LIMITLESS JOURNEY

ASTRONOMY'S LIMITLESS JOURNEY

A Guide to Understanding the Universe

Günther Hasinger

A Latitude 20 Book
University of Hawai'i Press
Honolulu

Translation assistance for some chapters provided by Orla Schmidt-Achert.

20 19 18 17 16 15   6 5 4 3 2 1

Library of Congress Cataloging-in-Publication Data

Hasinger, Günther, author, translator.
[Schicksal des Universums. English]
Astronomy's limitless journey : a guide to understanding the universe /
Günther Hasinger.
pages   cm
"A latitude 20 book."
Translation assistance for some chapters provided by Orla Schmidt-Achert.
Revised English translation of the author's Das Schicksal des Universums.
Includes bibliographical references and index.
Translated from the German.
ISBN 978-0-8248-5362-4 (paperback : alk. paper)
1. Cosmology.   I. Schmidt-Achert, Orla, translator.   II. Title.
QB981.H35613 2015
523.1—dc23   2015025628

Printed by Sheridan Books, Inc.

For Barbara, Oliver, Sebastian,
Caro, and the little one

Contents

Color plates follow pages 82 and 154.

Introduction

Over and over again, we are fascinated by the star-studded firmament. When looking up at it from a dark spot on a clear night, over six thousand stars are visible to the naked eye. Throughout the centuries, this was our cosmos, our "outer space." With every new scientific discovery, however, this cosmos has grown bigger. During the last century this realization process has moved along at a breathtaking pace. Today we know that even with the help of the biggest telescopes, we can only see a tiny section of the endlessly vast universe. Besides the fascination and awe one feels considering the size of the universe, many people are also overcome by fear in the face of its incomprehensible dimensions and the duration of its existence. In the end, we are driven by two questions: "Where did we come from?" and "Where are we going?" If the answers to these questions do not come from our current knowledge about space and time, however, we tend to be very suspicious about them or to even give up and crawl back into our intellectual shells. Albert Einstein, whose theories turned the classical concepts of space and time inside out, said that "common sense" was the sum of prejudices you gather during the first eighteen years of your life. We have come a long way since Einstein, but sometimes it seems the more we learn, the less we actually know. With every new discovery, we open another door to the unknown, and with every question that is answered, we find more doors—doors that are still locked.

On all sorts of different occasions, I enjoy giving lectures on the universe that everyone can understand. The audience's fascination and the mostly positive feedback encourage me to lecture on more topics. The long discussions after my presentation are very interesting and important to me, if only because I get a good idea of how an audience member looks at things and what he or she would still like to know. Often, the same questions come up, and it's usually the ones that are the hardest to answer because we just don't know the answers yet—and maybe never will. People ask, *What came before the Big Bang? What lies behind the edge of the universe? Where is the universe expanding to? Will it collapse sometime in the future? What is space? What is time? Are they infinite? What is inside a black hole? Are we alone in the universe?*

Sometimes, as a joke, I say to my audience at the beginning of a lecture, "You can ask me anything. Just don't ask me something I don't know the answer to!" Scientific findings are taking enormous steps forward, and many of the questions that we couldn't even ask a few years ago are being answered today. The answers to other questions—for example, about the life of the universe before its birth—are still being worked on.

Many audience members come to the podium after the lecture and express their frustration with subject matter they cannot comprehend. One of the most frequent criticisms is *What you're doing here is purely theoretical and doesn't have anything to do with reality. If I haven't seen something with my own eyes, I can't believe it. So why shouldn't I just go ahead and believe something else?* I try to explain to them that what we call "seeing with our own eyes" is exactly what astronomers do with their telescopes. The lenses in our eyes focus electromagnetic rays onto our retina (the detector), where they are then transferred into electrical signals and sent to our brain (the central computer). The brain interprets these signals by comparing them with the huge amount of stored information and models from past experiences to come to the conclusions "That's a tree. That's a face." Of course, the brain does make mistakes sometimes, which is clearly demonstrated by the well-known phenomenon of optical illusions. When the brain receives something it cannot assign to anything it already has, the information is stored and archived in the category of a "new experience." Reality as such doesn't exist. It is rather our brain that makes us believe in it by constantly comparing things. If you take our example and replace the eye's lens with the mirror in a telescope, the retina with a sensitive detector, and the brain's data processing with a big computer, you get a very similar situation to "seeing with your own eyes." Of course, all the astronomical results and pictures have to be analyzed and interpreted by the human brain. But would we ever dream of questioning the reality of an ultrasound picture, an X-ray, or what we see when we look through a microscope just because we can't see it with the naked eye?

Many other people ask me where they can find more information about what they've just heard. Many fascinating books have been written about the universe and the many breathtaking new discoveries. Several can be found in the appendix to this book. Famous scientists have written quite a few of them, and I have read most of them with the greatest pleasure. But nevertheless, everyone has his own special way of explaining complicated matters to an interested audience. That is why I am also frequently asked, *"Why don't you write a book yourself?"* And that's what I have done.

You can regard my book as a journey from the beginning of the universe with its incredibly hot Big Bang, more or less 13.8 billion years ago, up until its cold and dark end that lies in the faraway and distant future. In between lie the times in which extensive structures, galaxies, stars, and planets form. At least one of these planets has turned out to be quite a good place for creatures to live on, including human beings who ask what they are made of, where they came from, and where they are going. And when it comes to exploring the far distant past, astronomy and astrophysics can help them out.

The history of the latest discoveries in astronomy and astrophysics can only be described as astonishing. Thanks to new, powerful telescopes; rapid advancements in numerical astrophysics; faster computers; and more sophisticated algorithms and observation methods, we are now able to explore new fields of research. Astrophysics and cosmology are experiencing a golden age, in which fundamental changes of our understanding of space and time, the origin and the future of the universe, and the exploration of our cosmic sphere are taking place. The Hubble Space Telescope, observatories over twenty-five feet in diameter at the best ground-based observing sites, huge radio telescopes, and a wide array of satellites have led to scientific breakthroughs and paradigm shifts. These satellites and observatories have wavelength ranges that are not accessible from Earth's surface, such as the X-ray satellites ROSAT, Chandra, and XMM-Newton; the infrared satellites ISO, Spitzer, and Herschel; and the microwave observatories COBE, WMAP, and Planck, During the last decades, new discoveries have been raining down on us. Not only have we seen radical changes in cosmology, where we have a commonly accepted model of the origin and development of the universe, but we are constantly discovering new planets outside our solar system. To date we know of two thousand planets. A paradigm shift has also taken place concerning black holes, which have transformed from being pure figments of the imagination to actual celestial bodies that have a significant influence on the evolution of our universe. Even the mysterious gamma ray bursts could be identified in the last decades as the birthing processes of black holes.

This book presents the newest discoveries about our universe and how we found them. It examines the methods used in astrophysics and the human side of celestial exploration. We will see the areas where our knowledge is incomplete and our journey gets a little shaky. Needless to say, the boundaries between knowledge and ignorance are constantly shifting.

Because all cosmic signals take a certain amount of time to reach Earth, we can look at astronomy as a time machine: the farther away an object is to us, the younger it was when it emitted its light. So the deeper we look into the universe, the younger the phases of the cosmos become. It is almost like looking at the crib we slept in as a baby—or even the womb. Once we have a good idea of the shape of the universe to this point, we can risk exploring the future and making predictions. Naturally, the farther away these points in time are, the fuzzier the information about the cosmos's past and future becomes. As we do more research, build bigger telescopes and particle accelerators, and develop more sophisticated theories, however, everything becomes clearer. It also makes us realize how much we still have to learn. Thus, the entire history of the cosmos, including the story of this book, is part of the everlasting evolution.

Come with me on this limitless journey!

Chapter One

THE DARK SIDE OF THE UNIVERSE

In the beginning God created the heavens and the Earth. And the Earth was without form, and void; and darkness was upon the face of the deep. And the Spirit of God moved upon the face of the waters.

—*Genesis 1:1–2*

TOHUWABOHU: THE CHAOTIC BEGINNING

Amazingly, the account of the creation of the universe from the first book of Genesis in the Bible, apart from a few important details, is not very different from that imagined by astrophysicists and cosmologists today. The Hebrew term "Tohuwabohu" (hullabaloo), which was translated in the English version of the Bible as "formless and void," is now a synonym for chaos and confusion. Indeed, modern cosmology assumes that our universe was created in a chaotic inflationary phase from emptiness. If the most recent astrophysical ideas are correct, we need to familiarize ourselves with the idea that what we call "nothing" (i.e., the vacuum or empty space) is the highest energy state of the universe. In some places this "nothingness" is strained to the point that it ruptures, similar to a wound-up spring in a watch. In a figurative sense, one can imagine the Big Bang as being a result of such tension. It is not known, however, how this force, called the "inflaton field," contributed to the creation of the universe. Nor do we know if this force has anything to do with the "dark energy" discovered around the turn of the millennium. Dark energy is also causing an exponentially accelerated expansion of the universe. The

1

findings about this hullabaloo have a long and varied history, in which our own Milky Way plays an important role.

When we look at the sky on a dark, starry night, we can see the bright band of the Milky Way. Figure 1.1 shows a panorama of the entire sky. The bright band running across the firmament is the Milky Way. In the middle of the image is the Galactic Center, which is invisible to the naked eye because of the dark dust clouds that cover it. When Galileo first observed the Milky Way through a telescope in 1609, he could see that the faint glow consisted of thousands of stars. It seems we are inside a disk of billions of stars. Almost all of the objects we can see in the night sky without the use of telescopes are stars in the Milky Way—the Galaxy. In the lower right of Figure 1.1, we can see two small nebulae, which the Portuguese navigator Ferdinand Magellan discovered in 1521. Unlike astronomers one hundred years ago, we now know that the Magellanic Clouds are two small separate galaxies close to the Milky Way.

If we could look down from above the Milky Way, we would see that our Galaxy is a gigantic spiral of stars and nebulae. Figure 1.2 shows the beautiful spiral galaxy NGC 1232. The image of NGC 1232 was taken with the Very Large Telescope (VLT) at the European Southern Observatory in the Chilean Atacama Desert.[1] This photo was

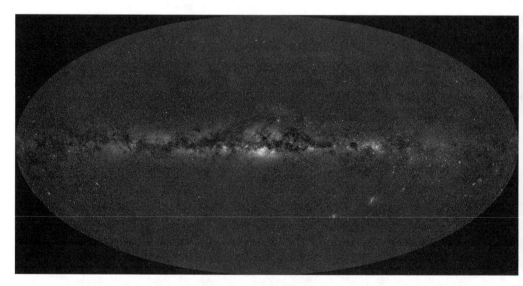

Figure 1.1. Panorama of the entire sky in visible light. In the middle of the figure is the Galactic Center. The galactic plane is visible as a bright patch streaked with dark clouds. (Courtesy of Axel Mellinger, Central Michigan University.)

Figure 1.2. The "first light" of the Very Large Telescope of the European Southern Observatory (ESO) on Cerro Paranal in Chile. The image shows the spiral galaxy NGC 1232, seen almost exactly from above. (Courtesy of ESO.)

taken almost exactly from above. We can see the young blue stars in the spiral arms of the galaxy that look like a string of pearls; in the center, the older, red stars dominate. The Milky Way and other typical spiral galaxies contain about one hundred billion stars. Our solar system is located between two of the outer spiral arms of the galaxy. When we look at the Milky Way, we are seeing stars, gases, and dust clouds.

For centuries the starry heavens above us have been regarded as immortal, only interrupted by the regular motion of the planets and sometimes by the ominous appearance of comets or "guest stars" that become visible for a few weeks. For most astronomical contemporaries of Albert Einstein, the band of the Milky Way alone defined the universe that had existed since the beginning of time and would continue into eternity. At that time, it was believed that the sun was at the center of the cosmos. The galactic band looked almost the same from all directions, it had the same number of stars on all sides, and the stars moved relatively slowly around one another.

When Albert Einstein developed the first version of the theory of general relativity in 1915, he realized that the solutions to his field equations did not allow a static, spatially bounded universe but one that would have to either collapse or break into pieces. We can just imagine the impact that this concept of an unstable, dynamic universe had on researchers at the time. Thus, in 1917, Einstein felt compelled to incorporate an additional stabilizing term into his equations, which he called the "cosmological constant Λ" (the Greek letter lambda). In a letter to his colleague Paul Ehrenfest, he said, "I have committed another crime in the theory of gravity, which puts me in slight danger of being imprisoned in a madhouse."[2] This constant can be looked at as kind of a repulsive effect—an antigravitational force. For example, if you throw a rock into the air, it will be pulled by gravity and fall back to the ground (of course, if you attach it to a rocket engine, it will be able to fly into space). If you were able to throw the rock a certain distance from Earth, however, the repulsive cosmological constant would make the stone move faster and fly away from Earth. Einstein therefore had to adjust the value of the cosmological constant Λ in such a way that this repulsive force precisely balanced the gravitational attraction of all cosmic matter and thus allowed a static universe. Still using our rock example, it would have to just come to rest at a certain distance from Earth. Today, however, we know that even with Einstein's cosmological constant, the universe would not really be static. The smallest disturbance would throw it off balance.

Even if you removed all matter and radiation from the cosmos—and from Einstein's equations—the cosmological constant would still remain. The "nothingness" would still be uniformly filled with a substance. Einstein and his scientific contemporaries were struggling with this problem for many years. Einstein himself, who was heavily influenced by the Austrian physicist and philosopher Ernst Mach, originally postulated the Mach principle, according to which inertial space does not exist by itself but is created by matter. In a 1916 letter to Karl Schwarzschild (whose name will come up again in our discussion of black holes), Einstein wrote, "You can put it jokingly like this: When I make all things disappear from the world, according to Newton the

Galilean inertial space remains, but in my opinion *nothing* will be left" (italics in original).[3] He believed the Mach principle would tell the difference between reasonable and unreasonable solutions to his field equations. Before Einstein published his theory of general relativity, however, Dutch astronomer (and later director of the observatory in Leiden) Willem de Sitter had published a solution to Einstein's field equations for a mass-free universe dominated by a cosmological constant. Einstein, of course, immediately criticized the solution of the gravitational equations proposed by his opponent. Only much later was it discovered that such a "de Sitter universe" would expand exponentially; this would become very important in the context of the inflationary theory described below. In 1921, Russian meteorologist Alexander Friedman published general solutions to the Einstein field equations that showed that the cosmos, with or without a cosmological constant, could expand or contract, depending on which parameters were used in the equations. Einstein, who met the dynamic solutions of his equations with great skepticism from the beginning, originally claimed to have found a calculation error in Friedman's solutions, but in 1923, he had to admit that Friedman was correct, which clearly shook his confidence in the Mach principle.[4]

General relativity originally had exclusively astronomical applications. It received its baptism of fire from Einstein's explanation of the hitherto unexplained astronomical observations of the precession of Mercury's orbit. The first measurement of the gravitational deflection of light at the solar limb, taken in 1919 by English Astronomer Royal Sir Arthur Eddington, made Einstein world famous overnight. Einstein, however, was initially less interested in the impact of these dramatic developments on his theory than, for example, his colleague de Sitter. Around 1920, a dispute culminated among astronomers that would usher in a new "Copernican revolution" in our understanding of the cosmos: the great debate about "island universes." At the end of the eighteenth century, Immanuel Kant tried to determine if some of the elliptical diffuse nebulae that can be seen with large telescopes consisted of independent, disk-like stellar systems—specifically island universes—or of glowing gas clouds in the Milky Way, such as the Great Nebula in Orion.

By 1899, astronomer Julius Scheiner had already recorded a spectrum of the Andromeda Nebula (M31) at the Astrophysical Observatory Potsdam and found that the light distribution of the nebula is very similar to that of the sun.[5] He concluded that the spiral nebula in the constellation Andromeda should be a system of stars like the Milky Way. He proposed that the Milky Way also had a spiral structure, a unique idea at the time. In a 1909 article, Scheiner complained about the director of his observatory, Hermann Carl Vogel, who had refused to publish Scheiner's ideas in *Popular Astronomy*.[6] In the same article, he also criticized astronomer Edward Arthur Fath, who doubted Scheiner's theory, even though he later arrived at the same conclusion from his own research. Scheiner was finally rehabilitated almost a century later when Hans Oleak of the Astrophysical Institute Potsdam managed to digitize Scheiner's almost completely faded original photographic plate on a modern machine in 1995 and could fully confirm his statements.[7]

In 1920, the famous "great debate" between US astronomers Harlow Shapley and Heber D. Curtis was held at the National Academy of Sciences in Washington, DC. The two astronomers argued about the nature of the nebulae. Curtis believed that the nebulae consisted of gas clouds in the Milky Way, whereas Shapley saw them as independent island universes made up of millions of stars. As it turned out, the proponents of both camps were right. One of the hardest problems in astronomy is determining the distance to celestial bodies, and at that time it was not possible to accurately measure the distance of the nebulae. With ever-improving telescopes, it finally became possible to resolve individual bright stars in some of the nebulae and thus determine their distance; some nebulae turned out to be gas clouds.

Completing a long-term study of a group of variable stars in the Magellanic Clouds, in 1912, US astronomer Henrietta Leavitt showed that their pulsation period is directly related to their luminosity. From this fact, one could develop a new technique for determining distances, called the "Cepheid method." By 1913, Danish astronomer Ejnar Hertzsprung, working in Potsdam, had estimated the distance to the Small Magellanic Cloud. Using the variable star Delta Cephei, after which the method is named, he determined that this nebula is a stellar system outside our Milky Way—an island world of its own. The distance he specified at the time—about three thousand light-years—however, was ridiculously small, probably due to a calculation error, and as a result, his groundbreaking work got very little attention internationally. The breakthrough, and the end of the island universe debate, came in 1929 with Edwin Powell Hubble. Using the largest telescope at the time—the 2.3-meter mirror on the Mount Wilson Observatory—Hubbell looked at individual Cepheid stars in the

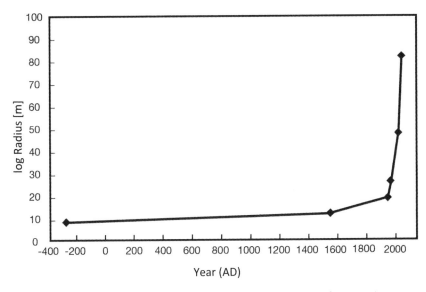

Figure 1.3. The size of the known universe as a function of time. Several "Copernican revolutions" have dramatically increased the known universe: Copernicus showed in 1500 that the sun, not Earth, is the center of the solar system. However, until 1920, astronomers believed that the sun was the center of the Milky Way. Only Shapley proved that the sun is located at the edge of our Milky Way, and in 1929, Hubble determined that our Milky Way is only one of many island worlds. The theory of inflation and the accelerated expansion of the universe have pushed the boundaries even farther. Judging from this curve, there is no end in sight.

Andromeda Nebula Messier-31 and discovered that the diffuse nebular spot was in fact a huge star system that was similar to the Milky Way but 2.1 million light-years away. Since that time, the term "extragalactic"—meaning located outside the Milky Way—has been used to describe this type of system. In addition, Shapley used the distance determination of globular clusters, also at Mount Wilson, and proved that the sun is not in fact located in the center of the Milky Way but is actually much closer to the edge of the Galactic disk. Thus, with one blow, the size of the universe had expanded many times over, and once again, as with Copernicus, mankind was exiled from the center of its universe into a much larger space. Figure 1.3 shows how information about the size of our universe has increased dramatically over time.

However, an even larger breakthrough is connected with the name of Edwin Hubble, and this brings us back to Einstein. Much like the police determine a car's rate of speed in a speed trap, the rate of movement of galaxies can also be measured. This is analogous to the effect that Austrian physicist Christian Doppler discovered in 1842: the frequency of sound or light waves varies with the speed of the object. You may have noticed that the sirens on police cars sound louder when they are approaching from behind than when they're driving away. The effect is even more pronounced in Formula-1 races where the audience hears the characteristic "wrrrooam" of the cars racing by. Just think of how it sounds to the race car driver himself!

If you divide the light from a star with a prism or a finely ruled grating into its rainbow colors, you can see dark lines in this so-called spectrum. These are Joseph von Fraunhofer's spectral lines, the fingerprints of individual chemical elements in the atmospheres of stars. By 1912, US astronomer Vesto M. Slipher, working at the Lowell Observatory in Arizona, had already discovered that in some galaxies, these spectral lines are shifted to the red part of the spectrum. By 1928, he had collected forty such "redshifts." Willem de Sitter was the first to interpret these redshifts by the Doppler effect, corresponding to recession velocities of several hundred kilometers per second, and he included them in his model of the expanding universe.

In 1924, astronomer Carl Wilhelm Wirtz combined Slipher's redshift determinations with distance estimates based on the size of the spiral nebulae and proposed a linear relationship between velocity and the distances of galaxies.[8] A year later, Swedish astronomer Knut Lundmark published an extensive account about the motion and distance of spiral nebulae, where he found a definite correlation between the apparent dimension and radial velocity of galaxies.[9] As discussed above, Edwin Hubble measured the distance to the Andromeda Nebula in 1924 with the help of the Cepheid method. In the next five years, he was able to measure the distance of eighteen galaxies and discovered an approximately linear relationship with redshift, similar to Wirtz. Along with his assistant Milton Humason, he published a much more detailed and convincing sample in 1931.[10] This revealed that almost all galaxies are moving away from us and that they move faster the farther away they are from us. Our neighboring galaxy, the Andromeda Nebula, is one of the few galaxies that is moving toward us at about 150 kilometers per second. That finding will play an important role at the end of this book.

The discovery of the expansion of the universe is a good example of how credit for astronomical discoveries is not always given to those who deserve it. In 1927, Georges-Henri Lemaître, a Jesuit priest, published the idea of an expanding universe that emerged from a "primordial atom." Lemaître, a student of Harlow Shapley, had thus introduced a kind of creation act into cosmological models. If you look at the famous Hubble diagram of 1929, it is difficult to establish a significant correlation between velocity and distances of the galaxies, so you can understand why Steven Weinberg believes credit is also due to Slipher, Wirtz, Humason, and Lemaître.[11] The conclusions from these observations, however, were colossal, including that the cosmos actually expands! Like the raisins in bread dough that expands during baking, all galaxies are moving away from one another and at a speed that increases the farther apart they are. This suggests that in the early days of the universe, all matter must have been very close together, and thus the Big Bang theory was born. The age of the universe can be estimated directly from the expansion rate: the faster the galaxies are flying apart, the shorter the time since the Big Bang. Unfortunately, the Hubble constant—that is, the slope of the correlation between velocity and redshift—originally turned out to be much too large, which made it appear that the universe was only two billion years old (today we know it is 13.8 billion years old).

Einstein must have been well aware of these stunning developments in astronomy in the years since 1917. But he was still convinced of its static cosmos, although in the meantime he had expressed some doubts about the Mach principle. In 1927, however, at the Solvay Congress in Brussels, he rejected the solutions for expanding universes and told Lemaître, "Vos calculs sont corrects, mais votre physique est abominable" (your calculations are correct, but your physics is terrible).[12] In 1931, on a trip to the United States, he visited the California Institute of Technology and the Mount Wilson Observatory, where Hubble and Humason had made their discoveries. It was only then that Einstein was convinced that with the recession velocity of the galaxies, his image of the static universe was destroyed once and for all. With the expansion of the universe, the original reason for introducing the cosmological constant had become obsolete. Almost immediately after his return to Berlin, he gave a lecture at the Prussian Academy of Sciences, where he announced that his assumption of a static universe had become untenable by Hubble's measurements and that his original equations—without the cosmological constant—would explain the expanding universe in a simpler way.

In 1932, he and his former opponent Willem de Sitter wrote an article about the expanding universe as a solution to his original field equations. In 1933, Einstein again traveled to the United States, this time to escape the Nazis, and he never returned to his homeland. Many years later he wrote, "If the Hubble expansion would already have been known at the time, when I had developed the theory of general relativity, the cosmological constant would never have been introduced."[13] George Gamow, an astrophysicist whom Einstein knew from Princeton (and we will see later in connection to the Big Bang), wrote the following in his autobiography:

> Thus Einstein's original gravity equation was correct and changing it was a mistake. Much later, when I was discussing cosmological problems with Einstein, he remarked that the introduction of the cosmological term was the biggest blunder he ever made in his life. But this blunder, rejected by Einstein, is still sometimes used by cosmologists even today, and the cosmological constant Λ rears its ugly head again and again and again.[14]

Apparently, Gamow couldn't stand the cosmological constant! Evil tongues claim that he was glad Einstein had made this remark, or he may have been tempted to invent it himself. Einstein certainly was never able to make the story about his "greatest blunder" go away.

The fact that without the cosmological constant, the universe turned out to be much younger than, for example, Earth and many stars already known by that time did not bother Einstein very much. He said that in astronomy, one must be careful with too large extrapolations in time (he was right). Some proponents of Λ, such as Eddington and Lemaître, however, continued to stress the importance of the cosmological constant, and in the 1960s and 1970s, it was reintroduced by some astrophysicists. In Germany, astronomer Wolfgang Priester from Bonn University was an ardent advocate of the cosmological constant. Its existence and exact value were controversial for decades, but everybody agreed that Λ had to be very small. Its energy density could not be much larger than a few hydrogen atoms per cubic meter. It was only shortly before the turn of the millennium that the astronomical evidence became so overwhelming that the existence of a kind of cosmological constant could finally be proven. Several independent measurements indicate that there must be something like a repulsive gravitational force on very large scales, which was henceforth referred to as "dark energy" and accepted by the majority of astronomers.

Does this have anything to do with the "chaos" from which the universe was created in the first place? We still do not have the answer to that question. In fact, both the cosmological inflation in the first split fractions of a second of the universe and the current accelerated expansion of the universe are based on a repulsive force that causes an exponential expansion of the cosmos. Frankly, the information we have today about both inflation and dark energy is far too sketchy to draw such conclusions. To deal with this question in more detail, however, we must first discuss the weird quantum fluctuations of the vacuum.

MAYONNAISE AND THE FORCE FROM NOWHERE

How can we believe that "nothingness" is filled with energy? What is a vacuum anyway? In classical physics, a vacuum is what is left when all matter particles are pumped out of a space and its walls are cooled down to absolute zero so no more radiation is present. Since the middle of last century, however, we know that the vacuum must be described by quantum mechanics. Einstein had already shown in his theories of relativity that space and time lose their commonsense form when you have extremely high speeds or very large masses. The space is then contracted or curved, time is warped, and clocks run slower. On the other hand, if you push to very small dimensions—for example, by dividing a meter rod or a time interval over and over again—time and space at the smallest dimensions lose their shape as well and must be described by the laws of quantum mechanics. There may very well be a kind of "flickering," foam-like structure, in which the particles of matter and light assume a waveform and can only be described by probability distributions.

Imagine you are leaving on a trip on Saturday afternoon and are impatiently waiting for the mail carrier to deliver your passport and visa (this happened to me once or twice). Although you have a rough idea of when the mail usually arrives, you don't know what possible obstacles the mail carrier might encounter and whether she'll deliver the mail in time for you to catch your plane. This situation is very similar to the ambiguity in physical descriptions of very small entities: the uncertainty principle introduced by Werner Heisenberg in 1927. According to this, the location and the velocity of a particle can never be determined simultaneously with absolute precision. The inaccuracy in the two variables is given by Max Planck's famous constant $\hbar$. The same uncertainty applies between the energy of a particle and the time interval at which you want to measure this energy. Within a very short time interval, a particle can have an extremely high energy without violating the uncertainty principle.

What does this mean for the vacuum? It turns out that a vacuum, as defined by classical physics, cannot exist. One can regard a classic vacuum as a particle with zero energy. If you were to measure its energy in an arbitrary time interval, it would always turn out exactly zero. But this is forbidden by quantum mechanics because all quantities, including zero, are always associated with a certain degree of fuzziness. We therefore have to assume that a vacuum contains fluctuations. We can imagine such a fluctuation, for example, as a pair of virtual particles, two photons or a particle and its antiparticle. For a very short time interval, these two borrow energy from nothing and exist together for a brief moment, after which they quickly disappear and give back the borrowed energy to the vacuum. As long as this time interval is short enough and the energy is small enough that together they satisfy the Heisenberg uncertainty principle, everything is fine.

Thus, quantum mechanics postulates vacuum fluctuations. According to this, a vacuum is always filled with energy. However, these fluctuations are not just an abstract figment of a physicist's imagination, but they have everyday consequences that are directly observable on microscopic scales. If you, for example, bring an electron inside an atom to an excited state, it will spontaneously drop to the ground state after a certain time and will emit a photon from this transition. The spontaneity here is actually caused by a vacuum fluctuation. The most intriguing consequence of vacuum fluctuations is the force from the Casimir effect.

In 1948, Dutch physicist Hendrik Casimir investigated why colloidal solutions, such as mayonnaise and paint, are so "sticky." Small, micron-sized particles that are mixed into these liquids are connected in a kind of grid matrix and give the substance its cohesiveness. Casimir realized that the forces between two molecules in this solution could be explained by vacuum fluctuations.[15] He then wondered what would happen if two mirrors facing each other were placed close together in a vacuum. According to the uncertainty principle, the vacuum normally contains fluctuations of any wavelength (i.e., for example, any photon energy). Between two closely spaced metal plates, however, only fluctuations with wavelengths less than the distance of the plates can fit. Because any arbitrary fluctuation can be generated outside the plates, according to Casimir there must be a net force that pushes the metal plates together. This is the same type of force that pushes on the molecules in the mayonnaise. This force is usually very small, but it is measurable. Although the Casimir effect was extremely difficult to measure with the methods of that time, experiments with two mirrors have since been successfully carried out at Philips Laboratory; the results proved Casimir's theory.[16]

The most accurate measurements were achieved in 2001 with an atomic force microscope that measured the effect of a 0.2-millimeter-diameter sphere on a metal plate to about 1 percent accuracy (Figure 1.4). Of course, a ball slammed against a metal plate is not exactly the same as Casimir's configuration of two parallel metal plates. Nevertheless, a similar effect can be observed, and the ball has the advantage that it does not have to be aligned exactly to the metal plate and thus can be measured more accurately. The Casimir effect increases substantially with smaller distances, and at less than one micrometer, it is the strongest force between two molecules. At a distance of ten nanometers—about one hundred times the size of a typical atom—the Casimir effect, for example, already produces the pressure of one atmosphere! The physicists and engineers who develop

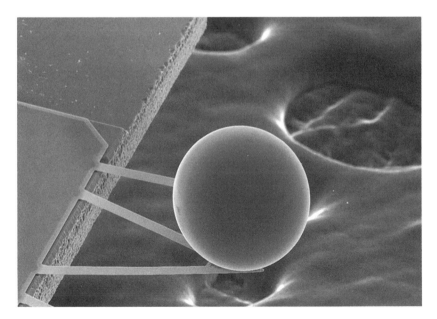

Figure 1.4. This tiny ball with a diameter of slightly more than one millimeter is pushed in the direction of the flat plate by quantum fluctuations. Umar Mohideen and his group performed the high-precision measurement of the Casimir effect from the University of California, Riverside. (Courtesy of Umar Mohideen, University of California, Riverside.)

tiny structures for the Nano-world—for example, Nano-robots that can swim around in human blood vessels—have to deal with the Casimir force on a daily basis. Actually, there are already micro-mechanical components that work on the basis of the Casimir force.

But what does this have to do with cosmology? The Casimir force is an effect of relativistic quantum electrodynamics—that is, the theory describing electromagnetic forces on the smallest scales. Although this theory is very well understood, it is difficult to calculate the total energy of the vacuum. This is because the Casimir force gets stronger as the distance decreases, so formally the vacuum energy of quantum electrodynamics is infinitely large. Depending on the shortest wavelengths used to perform the calculation, you get dramatically different values of 10^{36}–10^{70} joules for the energy in a one-cubic-meter vacuum. According to Einstein's formula $E = mc^2$, this corresponds to at least ten billion tons of matter per cubic centimeter of vacuum! This totally nonsensical value is many orders of magnitude larger than the value of the cosmological constant derived from astronomy, which corresponds to only a few atoms per cubic meter. In its early years, quantum theory raised many questions, so physicists initially did not care much about this problem. One exception was the Austrian physicist and Nobel Laureate Wolfgang Pauli, who in the early 1920s explored whether zero-point oscillations in vacuum energy could exert a gravitational force. He came to the conclusion that in this case the radius of the universe would "not even reach to the moon."[17]

This reveals a fundamental problem between the two most successful pillars of theoretical physics of the twentieth century: the theories of relativity and quantum mechanics. Both theories have been verified experimentally in great detail, and today, more than a hundred years after their introduction, both are the basis of technological developments of immense economic importance. Nobel Laureate Leon Lederman estimated that about 40 percent of the US gross national product is based on quantum mechanics—for example, transistors, microelectronics, computers, lasers, modern chemistry, biotechnology, and nanotechnology. The theory of relativity, on the other hand, is the prerequisite for the GPS satellite navigation and, for example, the successful German truck toll system, also a billion-dollar market. And yet, at their interface, the two theories do not agree. This applies to those cases where a strong gravitational force acts in a small space, such as the Big Bang and black holes. The theory of relativity deals with gravitational force and is based on a homogeneous space-time continuum. Quantum theory deals with the unified forces of electro-magnetic, weak, and strong interactions and at its smallest scales contains a foamy space-time structure shaped by random fluctuations. For the past several decades, legions of physicists have attempted to combine these two theories into a unified quantum theory of gravity, including Stephen Hawking in Cambridge, England, and the researchers at the Max Planck Institute for Gravitational Physics. Einstein worked on this problem for many years until his death. Although some findings are promising, the answer has yet to be determined. Therefore, many questions remain about the beginning of the universe and the nature of black holes, which is what makes this research so exciting.

DARK MATTER IN GALAXY CLUSTERS

Galaxies are often not alone in the sky but are arranged in groups and whole clusters. The Milky Way, along with the Andromeda Nebula, is a member of what Edwin Hubble labeled the "Local Group" in 1925. The Local Group is home to a total of about forty members, mostly dwarf galaxies such as the two Magellanic Clouds, and has a size of about five million light-years. The Andromeda Nebula is the largest galaxy of the Local Group and is about 2.3 million light-years from the Milky Way. To get an idea of this distance, if the galaxies were the size of quarters, they would be about one yard apart. The two galaxies are moving toward each other at a speed of about 150 kilometers per second and are likely to merge in about three to four billion years. The Local Group is in the vicinity of a much larger galaxy system: the Virgo Cluster in the constellation Virgo, which contains several hundred galaxies. The Local Group is moving at a speed of about 200 kilometers per second through space in the direction of the Virgo Cluster. It is also moving at a speed of about 460 kilometers per second toward a super-cluster of galaxies in the constellations Hydra and Centaurus. The Virgo Cluster could be a complex structure on the outskirts of the Hydra-Centaurus supercluster. One of the largest galaxy clusters with several thousand members is about 200 million light-years away in the constellation Coma Berenices and is called the Coma Cluster. Clusters are the largest contiguous stable structures in the cosmos.

In a figurative sense, the galaxy clusters may be looked at like mega-cities in the cosmos that have already incorporated many smaller cities and towns. On the other hand, the Milky Way and the Androm-eda galaxy are probably more like small towns. However, the main

difference with this static image is that these accumulations are traveling at very high speeds. To illustrate the magnitude of these velocities,
we can compare them with those in the solar system. Earth rotates at
the equator at a speed of about one-half kilometer per second. The moon
moves in its orbit around Earth at about one kilometer per second,
and Earth orbits around the sun at about thirty kilometers per second.
The prerequisite that the moon or Earth will not be thrown off course
is that the attraction of the central celestial body keeps the balance
with centrifugal force.

US astronomy professor Fritz Zwicky, who was born in Bulgaria
and raised in Switzerland, was the first to recognize that something
was wrong with the velocities of the galaxies in clusters. In 1933, he
studied the galaxies of the Coma Cluster with the 1.2-meter telescope
on Mount Palomar. Although Zwicky's personal life was somewhat
colorful,[18] his contributions to astrophysics are astounding. When he
determined the velocities of the galaxies in the peripheral regions of
the Coma Cluster, he noted that they were moving much too fast. Just
as in the solar system, the gravitational attraction of the matter in the
cluster must balance the centrifugal force acting on the galaxies in the
outskirts to prevent them from flying out of the cluster. If, however,
the total number of stars and their total mass are deduced from the
light of all the galaxies in the cluster, it is not nearly enough to keep
the galaxies in the cluster. Zwicky concluded that a strong, invisible
force is needed to hold the cluster together, which is about ten times
stronger than the gravitational forces of all the galaxies combined. He
called the perpetrator of this force the "missing mass." This term is
actually misleading because it is not the mass that is missing but the
light that such a mass of normal matter would have to emit or absorb.
Today, this additional invisible component is called dark matter. We
will look at some of Zwicky's other important discoveries later.

GALACTIC ROTATION CURVES

Also in the 1930s, the famous Dutch astronomer Jan Hendrik Oort
studied the velocity of the stars in the solar system. From the spatial
distribution of the brightest stars near the sun, he was able to estimate
the thickness of the galactic disk in the solar system to about two thousand light-years. He also found that, using the velocities of these stars,
he could deduce the mass that is necessary to keep the stars within the
disk. Like Zwicky, he came to the conclusion that the total mass of

stars and gas clouds in the solar neighborhood is insufficient to keep the stars in the Milky Way. He calculated that two to three times more than the mass of the visible stars is necessary for this.

For many decades, few astronomers cared about Oort's and Zwicky's missing mass. Only since the 1960s and 1970s have new, independent measurements proven that indeed a large fraction of matter in the universe must be dark. Comparing the velocity of the stars toward the center of the Milky Way with those moving in the opposite direction, it is found that the sun moves around the Galactic Center with a speed of 220 kilometers per second. Because the radius of the solar orbit is about twenty-five thousand light-years, it takes about 240 million years for one revolution. If you interpret the movement of the stars around the Galactic Center in the same way as the planets' orbits, using the laws established by Johannes Kepler early in the seventeenth century, you might expect that the stars in the inner region of the galaxy revolve around the center much faster than those in the outer regions. In the solar system, Earth orbits the sun in exactly one year, while Jupiter takes twenty-nine years, and Uranus takes more than eighty-four years. Mercury, on the other hand, takes only eighty-eight days to orbit around the sun. Because we are located inside the disk of the Milky Way, it is very difficult to determine the rotational velocities of stars within and outside the solar orbit. In 1965, Dutch astronomer Maarten Schmidt, however, derived a first rotation curve for the Milky Way from velocity measurements of hydrogen clouds in the radio band, which looked completely different from the curve predicted on the basis of Kepler's laws. Schmidt pointed out that large amounts of invisible matter must still exist outside the sun's orbit. We will see Maarten Schmidt in more detail later when we talk about quasars and black holes in the centers of galaxies.

Only US astronomer Vera Rubin's measurements of galaxies' rotation curves proved beyond a reasonable doubt that large amounts of dark matter existed in galaxies' outer regions. In 1970, Rubin and her colleague Kent Ford published a very comprehensive work on the rotation velocities in the Andromeda Nebula. Spectroscopic observations allowed her to measure the speed of many luminous gas clouds in our neighbor galaxy using the Doppler effect. Surprisingly, the derived rotation curve did not fall off toward the edge of the galaxy but was almost completely flat. The stars at the outer edge of the galaxy moved as fast around the center as the interior stars (Figure 1.5).[19] How could that be? What kept the galaxy from flying apart from centrifugal force?

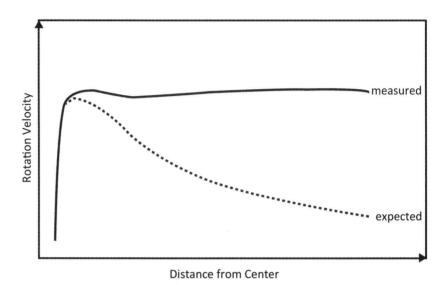

Figure 1.5. Schematic rotation curve of a galaxy. The solid line indicates the measured curve; the dotted line shows the curve expected from Kepler's laws without dark matter.

First, Rubin considered that the Andromeda Nebula was an exception. Along with Ford, she took spectroscopic measurements on several other spiral galaxies, but the results were always the same: flat rotation curves. Only then did she remember the undergraduate exercise involving Fritz Zwicky's "missing mass" in galaxy clusters and realized that her rotation curves probably pointed to the same dark matter that Zwicky had needed to hold the galaxy clusters together.

All galaxies must therefore be embedded in an extended cloud of dark matter, a so-called dark halo that extends far beyond the visible area of the luminous galaxies. In the case of the Milky Way and the Andromeda Nebula, it could well be that their dark halos touch, or even merge. The entire Local Group of galaxies is probably sitting in a common dark matter halo that reaches to the outer regions of the Virgo Cluster.

GRAVITATIONAL LENSES

The year 1905 was Einstein's *annus mirabilis*—his miraculous year. That year he published four trailblazing papers on the theories of special relativity ($E = mc^2$), the photoelectric effect (the Franck-Hertz experiment), and the Brownian molecular motion, as well as his dis-

sertation thesis. He began work on his theory of general relativity in
about 1907. In 1911, Emperor Franz Joseph appointed him as professor of theoretical physics at the German University in Prague. A year
later, he was appointed Professor of Theoretical Physics at the Swiss
Federal Institute of Technology in Zurich, and in 1914, he transferred
to the Royal Prussian Academy of Sciences in Berlin. During these
years he met the astronomer and assistant at the Berlin Observatory,
Erwin Finlay Freundlich. Founded in 1700, the traditional Berlin Observatory was transferred to the Babelsberg near Potsdam in 1913
because of the better observing conditions there. From the beginning,
Freundlich was impressed with the new theory was credited by Einstein
as "the first who has taken the trouble to subject the theory to a test."
Starting in 1911, Freundlich worked with Einstein on concepts for the
experimental verification of the theory of relativity—much to the cha-
grin of his boss, the former director of the Royal Observatory in Berlin,
Hermann Struve, who insisted that Freundlich should work on the
completion of the Potsdam sky survey atlas and strongly dissuaded him
from engaging with Einstein's revolutionary ideas.

One of the key predictions of Einstein's theory of relativity was
that the existence of a large mass bends space. A ray of light that
travels the shortest path through such a "dent" in space needs to be
deflected from a straight line. According to Einstein, the light rays of
stars behind the sun should be deflected by a tiny angle (about two
seconds of arc, or one part in two-thousandths of a degree). In 1913,
Freundlich published an article in the journal *Astronomische Nachrichten* titled "About an Experiment to Prove the Deflection of Light
in Gravitational Fields Suspected by A. Einstein."[20] Freundlich realized that this effect could only be observed during a total solar eclipse,
where the light of the sun would be reduced so much that the stars in
the background would be visible. In his article, he asked all astronomers
to photograph the upcoming solar eclipses. Due to lack of response,
Freundlich traveled to Russia himself to photograph the eclipse on August 21, 1914. The Prussian Academy and the Foundation of Gustav
Krupp von Bohlen und Halbach financed the trip. With a colleague
and a technician, he left Berlin on July 19 and arrived in Theodosia in
the Crimea peninsula a week later, where expeditions from Argentina
and the United States had already arrived. Freundlich, however, had
brought the most powerful telescopes and cameras. Unfortunately,
World War I broke out on July 28, and all plans and hopes were dashed.
Freundlich and his German colleagues were interned overnight as

enemy foreigners, and their equipment was confiscated. (Ironically, it would have been impossible to photograph the eclipse anyway because of heavy clouds. A member of the US expedition wrote in his diary, "Thick gray clouds during the eclipse darkening and wonderful sunshine afterwards.") Einstein expressed fears for his "good astronomer Freundlich," who had to "experience captivity in Russia instead of the eclipse." In late September, Freundlich and some other German captives were exchanged for some Russian officers and released.[21] Freundlich's equipment remained lost for many years while the war raged.

I am one of those hopeless optimists who, even in the face of a negative result or a bad experience, still ask, "Who knows what it is good for?" Einstein and Freundlich had originally used only the theory of special relativity to calculate the deflection of light. By 1915, they had the final equations of general relativity that also took into account the curvature of space and predicted the deflection of light at the solar limb to 1.7 arc seconds. This value is exactly twice as large as the one that arose from Newton's theory of gravity or the theory of special relativity. Therefore, maybe it was fortunate that none of the eclipse expeditions before 1916 were successful. Imagine if the 1912 and 1914 expeditions had measured a value twice as large as the one originally predicted by Einstein and then had to be corrected afterward!

The disruptions caused by war and bad weather ultimately helped English astronomers to obtain the crucial experimental confirmation of general relativity. During the war, it was almost impossible to distribute Einstein's work in enemy countries. But the young astronomy professor Arthur Eddington, who held the Plumian chair originally established by Isaac Newton at the University of Cambridge and was later appointed Astronomer Royal and knighted, had a private copy of Einstein's great article on the theory of general relativity from 1916. In the middle of the war, the London Admiralty absolved the latent conscientious objector Eddington from military service under the condition that he would arrange an expedition to a solar eclipse near the equator on May 29, 1919. This eclipse appeared particularly well suited because the sun would be in front of the bright star cluster of the Hyades, thus allowing particularly accurate measurements. Immediately after the war, Eddington therefore initiated an expedition to the island of Principe off West Africa, while a second English expedition was on its way to Brazil. After several weeks of nice weather in

West Africa, there were once again clouds and rain exactly on the day of the eclipse. Nevertheless, Eddington steadfastly carried out the planned observation and had a blessing in disguise: one of his sixteen photographs of the eclipse produced a useful measurement of the brightest stars in the vicinity of the sun. The expedition in Brazil was successful as well. After comparing measurements for several months, voyaging back by ship, and conducting the necessary evaluations, the British officially announced their results in London. By that time, however, the news had already spread around the world like wildfire: Eddington has confirmed general relativity, and Einstein was suddenly world famous.

This leads us to the subject of gravitational lenses. Like the effect of light deflection at the solar limb, Einstein had predicted in 1937 that a light source that is exactly behind a massive celestial object as viewed by the observer should be mapped into a circle of light.[22] The massive object thus acts as a "gravitational lens." The radius of this "Einstein ring" can be directly connected to the mass of the lens. But Einstein himself considered this effect to be much too small to be visible because he had only included stars as possible gravitational lenses in his considerations. Fritz Zwicky, however, pointed out that the gravitational lensing could indeed be quite considerable when it was caused by a large mass collection such as a galaxy or a cluster of galaxies. Just in time for Einstein's one hundredth birthday in 1979, the first lensed binary quasar was discovered. This is an extremely luminous object whose light is split into two images by an intervening galaxy. We will discuss this in more detail later.

If you are sitting by candlelight, sipping a glass of fine red wine and reading this book, you might notice that the refraction of light in the glass casts an interesting pattern on the pages—so-called caustics. The wine in the glass acts as a lens that focuses and distorts the light of the candle (Figure 1.6). In the same way, the total matter (ordinary matter and dark matter combined) in a cluster of galaxies acts as a gravitational lens and distorts the light of galaxies located behind the cluster into interesting, luminous arches, such as those visible in the beautiful Hubble Space Telescope image of the galaxy cluster Abell 2218 (Figure 1.7). From these bright, lensed arcs, one can immediately infer the mass of the intervening gravitating matter, which is many times larger than the mass of all the galaxies in the cluster. This result is one of the best confirmations of the existence of dark matter.

Figure 1.6. Caustics from a wineglass as a model for gravitational lenses.

Figure 1.7. Hubble Space Telescope image of the galaxy cluster Abell 2218. All luminous galaxies are members of the cluster—a total of several hundred. The large luminous arcs are images of galaxies lying far behind the cluster, distorted through the strong gravitational lensing effect. From the radius of the luminous arcs, which are approximately segments of an "Einstein ring," one can infer the mass of the dark matter present. (Courtesy of W. Couch, R. Ellis, NASA, and the European Space Agency.)

THE PARTICLES OF THE STANDARD MODEL

Before we turn to the possible candidates for dark matter, we must first try to understand the composition of normal matter or, according to Goethe, "what holds the world together in its inmost folds." The known elementary particles and the fundamental forces acting between them can be very well described by the standard model of particle physics—an elegant, although somewhat complicated, theoretical model that was confirmed experimentally in recent decades with amazing accuracy and has produced several Nobel prizes, including for the discovery of the Higgs particle in 2013. Elementary particles are the quasi point–like, not further divisible, fundamental building blocks of matter. The Greek scholar Democritus was the first to postulate that matter is constructed of minute, indivisible building blocks, which he called "atomos," after the Greek word for "indivisible." Nineteenth-century chemists used Democritus' ideas when they discovered that chemical reactions always proceed in constant mixing ratios. The atoms have this way been identified as the chemically indivisible building blocks of the elements of the Periodic Table. Toward the end of the nineteenth century, in connection with the study of radioactivity, it was determined, however, that atoms are made up of smaller building blocks, the elementary particles. The electron was the first elementary particle discovered.

In the beginning of the last century, atomic physicists investigated the phenomenon of radioactivity. Otto Hahn and Lise Meitner studied the different types of radioactive decay, which were called alpha, beta, and gamma decay. In alpha decay an "α particle"—a helium nucleus— is emitted. In beta decay a "β particle"—an electron—is emitted. In gamma decay a "γ particle"—a photon—is emitted. Exactly the same amount of energy should always be emitted when a certain type of atomic nucleus decays. While investigating the details of the beta decay, Meitner and her colleagues were surprised to find different amounts of energy from the emitted electrons in their lab. Niels Bohr even conjectured that this might be a violation of the energy conservation law. In 1930, the "Group of Radioactives" met at the "Gauvereinstagung" in Tübingen. The aforementioned Austrian physicist Wolfgang Pauli, now a professor at the Swiss Federal Institute of Technology Zurich, could not participate, but he wrote an open letter to the conference attendees:

Dear radioactive ladies and gentlemen,

As the bearer of these lines, whom I ask you to graciously listen to, will dispute with you in more detail, I had to . . . resort to a desperate way to save the exchange law of statistics and the energy law. Namely the possibility that electrically neutral particles, which I like to call *neutrons,* exist in the nucleus, which have spin ½ and obey the exclusion principle and also discriminate themselves from photons in that they do not move at the speed of light. . . . The continuous β-spectrum would be understood by assuming that in each β-decay a neutron is emitted in addition to the electron, such that the sum of the neutron and electron energies is constant. I currently do not dare to publish anything about this idea, and trustfully turn to you first, dear radioactives, with the question how it stands with the experimental proof of such a neutron, if it would have a similar or about ten times larger penetration compared to a γ-ray. . . . I admit that my way out might a priori seem rather unlikely. . . . But only he who dares wins. . . . So, dear radioactives, now check and judge.—Unfortunately, I cannot appear personally in Tübingen, since I'm indispensable here due to a ball held in Zurich in the night of 6 to 7th of December. Best regards. . . .

Your humble servant W. Pauli.[23]

The Radioactives noted, judged, and enthusiastically approved of the idea of Pauli dancing in Zurich. The "spin" that Pauli mentions is a quantum mechanical property of elementary particles that can be imagined as a rotation around its axis—either clockwise or counter-clockwise. Particles with a half-integer spin are called "fermions," after the Italian physicist Enrico Fermi. They follow the so-called Pauli exclusion principle, which he called in his letter the "exchange law of statistics." According to this, two fermions may not have exactly the same quantum mechanical properties, a law that will become very important in the treatment of white dwarfs and neutron stars later on. Particles with integer spin are called "bosons," named after the Indian physicist Satyendra Nath Bose. Pauli originally named the new postulated ghost particle a "neutron." The particle, which is called a neutron today and is the neutral brother of the proton discovered in 1919, was identified in 1932. To end the confusion, in 1934, Fermi called the particle introduced by Pauli a "neutrino." Neutrinos are extremely difficult to detect because they obey only weak interactions. Therefore, it wasn't until 1956 that US physicists Frederick Reines and Clyde Cowen discovered neutrinos with an apparatus called a Poltergeist. This detector was built near a very strong nuclear reactor at Savannah River, South Carolina, and consisted of a tank with a cadmium chloride solution. The interaction of a neutrino with an atom in the tank was detected through the resulting characteristic gamma rays. Poltergeist discovered only about three potential neutrino events per hour. However, when the strong reactor beam was turned off, the signal went to zero. Reines received the Nobel Prize in 1995 for this measurement.

Although it was believed for decades that protons and neutrons were the smallest indivisible building blocks of matter, in the 1960s, it was discovered that protons and neutrons are in turn composed of even smaller particles, called "quarks." Gradually, the contours of the generally accepted standard model of elementary particles and fundamental forces crystallized. According to the standard model, there are four fundamental forces: the strong interaction or nuclear force that keeps the atomic nuclei together; the electromagnetic interaction, which causes all of the effects of electricity and magnetism, as well as light; the weak interaction responsible for, among other things, ghost particles such as neutrinos; and the gravitational force that pulls together the planets, stars, and universe on large scales.

In the standard model, the two types of elementary particles are matter particles and exchange particles that transmit the fundamental forces. The matter particles are fermions—that is, particles with a half-integer spin that are divided into so-called leptons (from the Greek word for "light") and the quarks. Among the leptons, we distinguish among the neutrinos, which have zero electric charge and are considered massless, and the electron, the muon, and the tauon, which have an electric charge of negative one and different masses. The three different families of neutrinos are the electron, the muon, and the tau. Thus, in total there are six different leptons. While the charged leptons participate in both the electromagnetic and weak interactions, neutrinos can react only via the weak interaction. The quarks, on the other hand, react in particular with both strong and weak forces and the electromagnetic interaction. The quarks carry fractional electrical charges of $-1/3$ and $+1/3$. Like leptons, there are six different types of quarks, which have colorful names like *up, down, charm, strange, top,* and *bottom.* In addition to the normal matter, there is also the corresponding antimatter—for example, the positron as a counterpart to the electron, antineutrino, and antiquarks for each matter particle.

The force-carrying exchange particles are bosons. For each of the four fundamental interactions, there is an exchange particle. The electromagnetic forces are transmitted by photons. For the weak interaction, there are three exchange particles, called W+, W–, and Z–bosons. The strong interaction is transported through the so-called gluons, of which there are eight different types. The very entertaining book *From the Big Bang to the Disintegration,* by physicist Harald Fritzsch, provides a compelling explanation of this process. Fritzsch was involved in the formulation of the quark theory, known as quantum chromodynamics.[24] The exchange particles in the electromagnetic, weak, and strong interactions each have spin 1, and all were confirmed in particle accelerator experiments in recent decades. According to the same theory, the gravitational force is transmitted by the "graviton," a particle with spin 2, which has not yet been discovered. Another particle of the standard model that had to wait for several decades to be discovered is the so-called Higgs boson, which is responsible for the generation of all the particle masses. In 2012, about two thousand scientists working on two different experiments at the most powerful particle accelerator, the Large Hadron Collider at CERN in Switzerland, finally announced the discovery of the Higgs boson, the last missing puzzle piece of the standard model. Theoretical physicists Peter Higgs and François Englert, who independently developed the key theory, were awarded the 2013 Nobel Prize in Physics.

At low temperatures, quarks cannot exist independently. They are glued together by the strong attraction of the gluon particles to form either a group of three quarks, the so-called baryons, or a quark and an antiquark, a so-called "meson," which is very short-lived.[25] The term "baryon" is derived from the Greek word *barys,* meaning "heavy," and indicates that these particles have a large mass in contrast to the light leptons and the medium-weight mesons. The main building blocks of matter, the protons and neutrons that build the periodic table of elements, are made of "up" (u) and "down" (d) quarks. The proton consists of two up and one down quark (uud) and the neutron of one up and two down quarks (udd). As we will see later, the baryons have been left over after gigantic amounts of quarks and antiquarks have annihilated one another in a ball of fire in the first second after the Big Bang. This process is known as "baryogenesis"—the creation of the baryons. The other, heavier quarks form a whole zoo of exotic particles that are very short-lived and therefore are of no concern to the fate of the universe.

THE PARTICLES OF DARK MATTER

Unlike the particles in the standard model, we currently know nothing about the nature of dark matter. All we know is that it cannot involve baryons. According to many physicists (and I concur), it must be an as yet undiscovered type of elementary particle. After a lecture at the North Rhine–Westphalia Academy of Sciences in June 2004, I made a bet with Karl Menten at the Max Planck Institute for Radio Astronomy in Bonn that if dark matter particles were not discovered by June 2014, I would give him a bottle of champagne. Obviously, this date has passed, and dark matter particles have not been discovered, so I have to ask Karl what kind of champagne he likes. Nevertheless, to motivate my original optimism in a positive outcome of this bet, I would like to discuss the recent discoveries about the neutrino ghost particles and the experimental searches for dark matter particles.

According to recent findings, neutrinos contribute little to dark matter. Pauli originally postulated that neutrinos weighed as much as electrons and moved more slowly than the speed of light. However, in the context of the standard model of particle physics, scientists took it for granted for decades that neutrinos were massless and therefore moved with the speed of light. Neutrinos are produced in large numbers in the nuclear fusion processes in the center of the sun. This is why in the 1960s, Raymond Davis and his colleagues began constructing a detector for measuring solar neutrinos in the Homestake Mine in the United States. The sun's internal structure and fusion process are very well understood, so the model accurately predicts the solar neutrino flux. After many years of measurements, however, Davis and his colleagues found only one-third of the expected neutrinos in their experiment. For a long time, it was therefore unclear whether there was something wrong with the equipment, with the solar model, or with the standard model of elementary particles. The neutrinos detected by Davis' device originated from a relatively rare side reaction in the solar nuclear fusion chain. It was therefore a sensation when in the beginning of the 1990s, several detectors, including the GALLEX experiment and its successor GNO, which were conducted under German management, could for the first time detect neutrinos from the main branch of the hydrogen fusion cycle and in turn found significantly fewer events than theoretically expected, so Davis' measurements were fully confirmed.

The detection of solar neutrinos by Davis, together with the discovery of neutrinos from supernova 1987A in the Japanese Kamiokande detector by Masatoshi Koshiba, was rewarded with part of the Nobel Prize in 2002. Today it is believed that neutrinos have a finite but very small rest mass, in contrast to the assumptions of the standard model. The experiments of the Japanese Super-Kamiokande detector point in the same direction. If the neutrinos have a mass, the particles of the three different families can transform into one another—a process called "neutrino oscillations." The missing electron neutrinos from the sun must therefore have been converted into neutrinos of another variety on the way to Earth. The evidence for neutrino oscillations was a revolution in particle physics because, for the first time, significant differences from the standard model were detected. This has important consequences for physics.

The neutrinos in the universe for the most part originate from the hot primordial soup, which was formed immediately after the Big Bang. As we shall see, there are about a billion times more neutrinos than, for example, protons and electrons. Therefore, even a very small neutrino mass can contribute a significant portion to the total energy of the cosmos. Current estimates suggest that the neutrinos add 0.2 to 1.9 percent to the total mass/energy density of the universe. This corresponds approximately to the total contribution of stars in the universe (Figure 1.8).

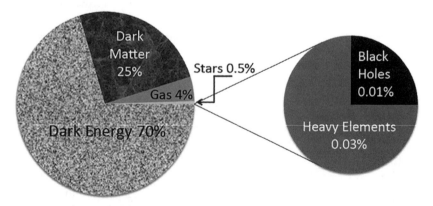

Figure 1.8. Contribution of the various known forms of matter and energy density in the universe (see also Table 1 in the Appendix).

Thus, neutrinos constitute part of dark matter, although far too little to explain the entire phenomenon. In addition, because of their low mass, the neutrinos move nearly at the speed of light. They are therefore referred to as "hot dark matter." As we will see in Chapter 4, however, we need "cold dark matter" to understand the development of large-scale cosmic structures. Dark matter particles must move at much lower speeds, and therefore have larger mass, to allow the large-scale cosmic structures to form.

For a certain time, the so-called MACHOs have received a lot of attention as possible candidates to explain the missing mass. MACHO stands for Massive Compact Halo Object—that is, heavy, invisible objects in the halos of galaxies. They could, for example, be black holes; brown dwarfs; or cool, and thus dark, white dwarfs. At any rate, they would be objects made from normal, baryonic matter. If there really are a large number of such objects, they must betray themselves by the gravitational lens effect discussed above. If one observes a very densely populated star field for a long time, it may happen that such a dark object passes right in front of a star and thus amplifies its light through the gravitational lens effect. This phenomenon is referred to as a "microlensing" effect and leads to a clear, very symmetrical rise and fall of the brightness of the star over several days.

A particularly interesting pattern of brightness enhancement occurs when one or more planets in the star system also cross the line of sight. In fact, a number of these microlens events were discovered in large observing campaigns by several different teams looking in the direction of the Galactic Center, the Large Magellanic Cloud, and the Andromeda Nebula. In the meantime more than thirty extrasolar planets have been discovered through microlensing. However, the number of the MACHO events is by far too low to contribute significantly to dark matter. The primordial synthesis of the light elements in the early universe, which is discussed in Chapter 3, leads to the conclusion that normal, baryonic matter can only contribute a maximum of 5 percent to the mass-energy balance of the cosmos. MACHOs are therefore practically excluded as candidates for dark matter, which amounts to 24 percent of the total mass/energy content of the cosmos (see Figure 1.8).

A very good candidate for cold dark matter, however, is a WIMP, which stands for "weakly interacting massive particle." (The acronym is modeled after the English term for "weakling," as opposed to a MACHO.) The designation indicates that this particle, like, for example, the neutrino, is only subject to the weak interaction and gravitation but with a much greater mass than the neutrino. Like neutrinos, WIMPs can react with normal nuclei through a weak interaction, such as an inverse beta decay, and could be detected in this way. To this end, several detectors, including the experimental CRESST (cryogenic rare event search with superconducting thermometers) that is used by the Max Planck Institute for Physics, Technical University Munich, and others in the tunnel under the Gran Sasso mountain range; the DAMA/LIBRA, also in the Gran Sasso Tunnel; and the EDELWEISS (expérience pour détecter les WIMPs en site souterrain) experiment that is used in the Frejus Tunnel in France, are in operation right now. In an old mine in Minnesota, dark matter detectors like CoGeNT (coherent germanium neutrino technology) and CDMS II (cryogenic dark matter search), which have produced some early tantalizing claims about the possible detection of a relatively low-mass WIMP, are in operation. However, the most sensitive direct dark matter searches performed by the Large Underground Xenon (LUX) experiment in 2013 and 2014 in the Homestead Mine in South Dakota have completely ruled out these results.[26] We therefore are still pretty much empty-handed when it comes to the discovery of a dark matter particle, so I owe Karl that bottle of champagne.

The WIMPs, which have been postulated as dark matter candidates, could fit very well into a modern extension of the standard model of elementary particles—namely, the supersymmetry (SUSY). According to this, each of the currently known elementary particles has a supersymmetric partner, which is designated by a small "s" before the name, such as the "selectron" or the "squarks." If the elementary particle has a half-integer spin value, and thus represents a fermion, its supersymmetric partner has an integer spin and is therefore a boson. Most of these particles should have an extremely short life span. The SUSY theory, however, states that the lightest supersymmetric particle should be stable and thus be a good candidate for the cold dark matter. Because none of these SUSY particles has been discovered so far, it is believed that their energies are far above the masses, which can be reached with the existing generation of accelerators. Even the European superaccelerator Large Hadron Collider (LHC) at CERN in Geneva has not yet detected any signs of supersymmetry (and therefore didn't help me win my bet either). But there are still hopes that the LHC could discover a whole zoo of supersymmetric particles when it starts measuring with its full energy potential in its second season after 2016.

Another dark matter candidate is the so-called axion. These particles have been proposed as solutions to another mystery of elementary particle physics: the lack of a dipole moment in a neutron. As we discussed above, neutrons contain positively and negatively charged quarks, and experiments show that the electric charge is actually widely distributed within the neutron. However, the centers of positive and negative charges always fall exactly on top of each other so that the electric dipole moment vanishes. In the context of the current theory of strong interactions, this effect is not fully understood to date, but it could be explained by the introduction of a new particle: the axion. The axion has no electric charge or spin, and it reacts with normal matter only through a weak interaction. Because in the fireball of the Big Bang, however, all particle species must have been created in approximately the same number, then as many axions as, for example, neutrinos should have been formed, so the axion is also a good candidate for dark matter. If axions exist, the sun should be a strong source of them. Inside the sun, hydrogen is fused into helium at about 17 million K, so gigantic amounts of X-rays are produced. As suggested by Ukrainian physicist Henry Primakoff, a tiny fraction of the X-ray photons can be transformed into axions in the strong electric field of the hot solar plasma. Like the solar neutrinos, these solar axions can travel freely to Earth and could there be converted back into X-rays through what is called the "inverse Primakoff effect" in a strong magnetic field. Because the probability of this process increases with the square of the magnetic field strength and the length of the magnet, however, an extremely strong and very large magnet is needed.

Coincidentally, a superconducting test magnet that was originally built for technology development of the LHC is no longer used at CERN. This linear solenoid is ten meters long and is one of the strongest magnets in the world. The experimental CAST (CERN Axion Solar Telescope)[27] was built as part of a large international collaboration in 2003. To find the X-ray photons converted from solar axions, X-ray detectors are installed at both ends of the magnet. The most sensitive equipment has been developed by the Max Planck Institutes, which include the Institute for Physics in Munich and the Institute for Extraterrestrial Physics (MPE) in Garching. This X-ray telescope was developed by Carl Zeiss for the X-ray satellite project ABRIXAS, along with a sensitive X-ray detector that was developed by MPE for the ESA XMM-Newton mission. Since 2003, the CAST helioscope has been used to search for axions and in the meantime has ruled out broad ranges of possible parameters for axions.

I still feel, more than ten years after my bet, that WIMPs are currently the best candidates for dark matter. Unfortunately, neither the eagerly awaited results from the LHC at CERN nor the direct detection attempts around the world have revealed any glimpse of supersymmetric particles or WIMPS. Therefore, the axions are still in the running. My hope to find dark matter particles in the near future has waned somewhat.

Chapter Two

THE BIG BANG

THE CRITICAL DENSITY

Now that we have come to know the main actors of the universe—dark energy, dark matter, and baryons—we have enough information to approach the complex subject of the origin of the universe. As Edwin Hubble and his contemporaries have shown, all of the galaxies are moving away from one another, and the farther away they are, the faster they travel. A straight line whose slope is the famous Hubble constant H_0 can describe the relationship between velocity and distance. This quantity has a very checkered history, especially because distances in astronomy are so difficult to measure. The generally accepted value is now around seventy-five kilometers per second per megaparsec[1] while Hubble's original value was about eight times higher.

If I throw a stone into the air, it rises. Due to Earth's gravity, it slows as it ascends and then pauses, only to fall back to the ground. But if I endow the stone with a sufficiently high speed—for example, by using a rocket engine—it can overcome its attraction to Earth and fly into space. Occasionally, large meteorites strike the surface of Mars and accelerate its rocks to such high speeds that they leave our neighboring planet and even reach Earth. This critical speed, called the "escape velocity," depends only on the mass of each planet.

We can imagine the motion of the galaxies in the universe in a similar manner. All the galaxies fly apart like the sparkling explosions of Fourth of July fireworks. But similar to the stone example, their

joint gravity should slow them down. Whether they halt and reverse
at some point in the distant future or whether they fly farther apart
for all eternity depends on the total mass of all galaxies and whether
their speed is greater or smaller than the corresponding escape velocity. From the known velocities of the galaxies, one can easily calculate
the critical density ρ_c required to stop the expansion of the universe.
It only depends on the Hubble constant[2] and corresponds to the tiny
value of about six hydrogen atoms per cubic meter. The critical density is incredibly small. For example, the best laboratory-made ultra-
high vacuum has a residual density of about 10^4 molecules per cubic
centimeter, and even the eternal emptiness of our Milky Way has a
density of about one hydrogen atom per cubic centimeter, which is
160,000 times higher than the critical density. This low value can only
be understood when you spread all the stars, galaxies, and clusters of
galaxies evenly across its almost-empty expanse. The real density of
the universe ρ is usually expressed in units of the critical density ρ_c and
is denoted by the Greek letter omega: $\Omega = \rho/\rho c$. We distinguish between the matter density Ω_m, which exerts an attractive force, and the
density of dark energy Ω_Λ, which has a repulsive effect.

We already know that according to Einstein's theory of general
relativity, matter bends space. Therefore, the density of the universe
in relation to its critical density also determines the geometry of space.
If such density is exactly equal to the critical density ($\Omega = 1$), we live in
a flat universe that follows the Euclidean geometry. Here, the sum of
the angles of a triangle is 180 degrees, parallel lines never intersect,
and the circumference of a circle relative to its diameter is given by
the number π. If the density is greater than the critical density ($\Omega > 1$),
the curvature of space is positive. In this case, the sum of the angles of
a triangle equals more than 180 degrees, parallel lines intersect, and
the circumference of a circle is less than π times its diameter. The surface of a globe is a good two-dimensional analog of such a space. If
you were to shine a beam of light into this curved space, it would eventually, after a very long time, arrive back at the starting point. Finally,
if the density is less than the critical density ($\Omega < 1$), space is negatively
curved. The surface of a saddle provides a two-dimensional analog of
this space. The sum of the angles of a triangle is less than 180 degrees,
parallel lines are separating farther and farther, and the circumference
of a circle is larger than π times its diameter.

Another fundamentally important quantity can be derived from
the Hubble constant—namely, the age of the universe. If, for simplic-

ity's sake, we first assume that the rate of expansion has remained constant throughout the history of the universe, we can easily extrapolate back to the time when all galaxies emerged from the same point, as in the case of the Fourth of July fireworks. This so-called Hubble time, about fourteen billion years, is already a good approximation of the age of the universe. Figure 2.1 shows this case as a linear expansion of the universe. We can compare this age with the age of the sun (4.55 billion years) and the age of the oldest stars in our Milky Way (about thirteen billion years). In reality, however, we must take into account that the velocity of the galaxies changes over time. If the universe contains five times the critical density in the form of matter ($\Omega_m = 5$), the expansion would slow down: the galaxies moved faster in earlier times. As the figure shows, in this case the universe would only be about seven billion years old—much younger than the oldest-known stars. Conversely, when repulsive dark energy dominates, it accelerates

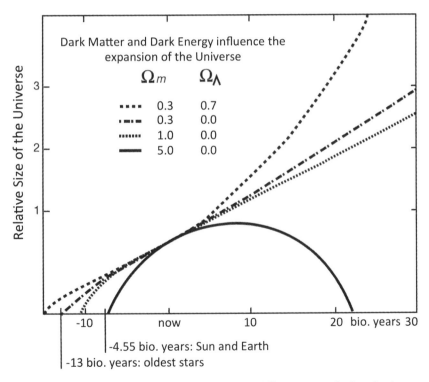

Figure 2.1. Different patterns of cosmic expansion for different assumed values for the energy density of matter (Ω_m is the sum of dark matter and normal matter) and dark energy Ω_Λ (bio. years stands for billion years). (Courtesy of NASA.)

the galaxies. They moved more slowly during earlier times, thus making the universe older than the Hubble time. The best currently available measurements, which we will discuss in the next few chapters, suggest that matter makes up about 30 percent of the critical density in the universe ($\Omega_m = 0.3$), and dark energy makes up about 70 percent ($\Omega_\Lambda = 0.7$). This corresponds to the top curve in Figure 2.1 with an age of 13.8 billion years. In the beginning, when the universe was much smaller and therefore matter dominated, the galaxies moved more slowly. Today, dark energy dominates, and the galaxies are accelerating. Therefore, the energy of the vacuum, the "nothingness," dominates the universe. Interestingly, the sum of matter density and energy density approaches the critical density ($\Omega_m + \Omega_\Lambda = 1 \pm 0.01$) and thus the geometry of the universe is almost exactly flat.

If all of this is true, and Einstein's cosmological constant actually dominates the energy balance of the cosmos, then according to Willem de Sitter's solution of Einstein's field equations, the universe is beginning an exponential inflationary phase. This could, in principle, go on in perpetuity, excluding deceleration and reversal of the cosmic expansion together with a reverse Big Bang. If it survives long enough, the matter in the universe will grow more and more diluted over the eons. The light will slowly grow weaker, and the temperature will drop farther and farther, until all eventually becomes terribly cold. The last chapter of this book further illuminates this possibility. However, we must realize that we still know nothing about the nature of dark energy or why it has approximately the same density as dark matter. Why do they not differ by a factor of 100 or even 10^{120}, as set out in Chapter 1? If dark energy is not the cosmological constant Einstein postulated—a constant energy that fills the vacuum, even if it contains no universe but is only a more complicated force—then the fate of the universe is left open again. One possibility for such a substance is the "quintessence," named after the fifth element sought by the ancient Greeks. Such a quintessence would not remain constant over time but could change in the future and even assume an attractive action. Future measurements with more sensitive, complete surveys in all possible observation windows are the only way to learn more about dark energy.

We must also gain a deeper theoretical understanding of the new physical questions, particularly regarding the unification of the quantum and relativity theories, to hopefully better explain both dark matter and dark energy. The current situation resembles the phase in physics more than one hundred years ago when extremely accurate measurements clearly disproved the classical ideas. Scientists all over the world searched for a new, unknown substance—the "ether"—in which light waves supposedly traveled. But not until Einstein presented his theory of special relativity, which questioned the concepts of space and time themselves, was ether thrown out of the race. Maybe in a few years or decades, we will experience the same course of events with dark energy. We may, however, need a female scientist named "Zweistein"[3] for this to happen.

One promising quantum theory of gravity, called "string theory," postulates that all known elementary particles are harmonic oscillations of the same string-like structures. These structures exist in ten or eleven dimensions, six or seven of which are "compactified" to very tiny scales so that only the four known dimensions, three of space and one of time, remain in our everyday world. String theory is very attractive but also very complicated. Even more than thirty years after its introduction, there still remains much to construct. More than a decade ago, I had the pleasure of participating in a faculty search committee for a string theorist and heard lectures about string theory from several excellent young physicists. After the fifth presentation, I thought I had at least a faint idea of what it was all about. I distinctly recall that all candidates believed that string theory would no longer be tenable should the existence of dark energy, discovered at about the same time, prove to be true.

But theoreticians are very inventive. Shortly afterward, a team around Paul Steinhardt of Princeton University developed an interesting variant of string theory, called the "ekpyrotic universe," to explain the cosmic acceleration of galaxies without invoking dark energy. In this concept, two four-dimensional "membranes" hover closely together side by side in a five-dimensional space. One of these membranes is our visible, four-dimensional universe; another is a parallel universe connected with ours only through gravitational force. The galaxies in our universe feel this gravitational pull and therefore accelerate. The two membranes float slowly toward each other and eventually intersect, freeing energy that corresponds to the hot initial state of our cosmos. Because this Big Splat can happen several times in the history of the universe, cyclic cosmologies are quite possible. The problem with string theories in general, however, is that they have so many possible incarnations. This quite poetic description of the ekpyrotic universe represents only one out of about 10^{500}–10^{1500} different possibilities. How can anyone find his or her way around there?

In terms of new theoretical concepts, I am usually very conservative, often just simply lacking the understanding. But sometimes it happens that over the years, more and more persuasive evidence is found for a new theory until even the skeptics slowly start to believe it. This has happened to me regarding the concepts of dark matter and dark energy, which have received observational support from several independent methods over the last decade. For the remainder of this book, I would therefore like to stay with the model of dark energy as a cosmological constant. An important piece of the puzzle, which I originally regarded as rather bizarre but which a growing number of observational results have confirmed, is the theory of inflation.

INFLATION: THE UNIVERSE TAKES A LOAN

Modern cosmology is a construction site. On one hand, astrophysicists try to fathom the past and the Big Bang by using ever more sensitive and comprehensive observational methods, thereby examining the largest structures in the universe, while on the other hand, theoretical and experimental particle physicists seek to improve the fundamental understanding of the universe's smallest particles and forces. In 1979, US physicist Alan Guth and Russian physicist Alexei Starobinsky both separately developed inflation theory to address problems with the Grand Unified Theory (GUT). Inflation theory later turned out to be extremely successful when applied to cosmology. Guth described the history of the theory of inflation in a beautiful book published in 1998.[4] I got to know the highly intelligent and slightly eccentric Starobinsky during my time as director of the Astrophysical Institute Potsdam. Starobinsky's model was rather complicated, and he did not give it the snappy name "inflation." Unfortunately, it was not widely distributed beyond the borders of Russia and East Germany and remained rather unnoticed in the West due to the Cold War and the limited travel options of that time.

We know the four fundamental forces: strong interaction, electromagnetic interaction, weak interaction, and gravitational force. The relative strength of each of these forces differs, in the ratio of $1:1/137:5\cdot10^{-14}:2\cdot10^{-39}$. Gravitational force is therefore almost forty orders of magnitude smaller than nuclear force. GUT theories assume that at extremely high energies, which prevailed in the very early universe and which particle accelerators on Earth will never reach, these four forces must have originated from a single elemental force (see Figure 2.2).

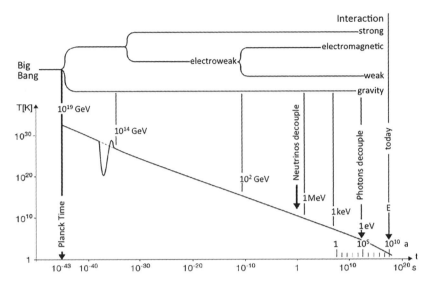

Figure 2.2. The entire history of the universe at one glance. The temperature of the universe is shown on an extremely compressed logarithmic time scale. The upper part of the figure shows the equivalent energy scale at times, corresponding to phase transitions in the splitting of the natural forces from a single elemental force.

According to the laws of quantum mechanics, there is a minimum length below which today's theories can no longer meaningfully describe physical processes. This Planck length is about 10^{-35} meters. Each object smaller than the Planck length would have a mass greater than the Planck mass of twenty-two micrograms and would be a black hole (see Chapter 8). This is where general relativity and quantum mechanics meet. Similarly, a minimum interval, a Planck time of about 10^{-43} seconds, is required for a light ray to pass through the Planck length. Therefore, the Planck time provides the earliest moment in which the current laws of physics can describe the universe (see Figure 2.2). At the Planck time at an energy scale of 10^{19} gigaelectronvolts, the gravitational force must already have separated from the other three fundamental combined forces, the GUT force. We can visualize this energy by comparing it with the energy we use in our daily life. An energy of 10^{19} gigaelectronvolts can light up a one-hundred-watt bulb for about eight months. Expressed as kinetic energy, this corresponds to a one-metric-ton car crashing into a wall at a speed of 220 kilometers per hour. However, the energy is concentrated in a tiny space—the size of a Planck length—leading to an incredibly high energy density. The twentieth-century quantum theories proposed by

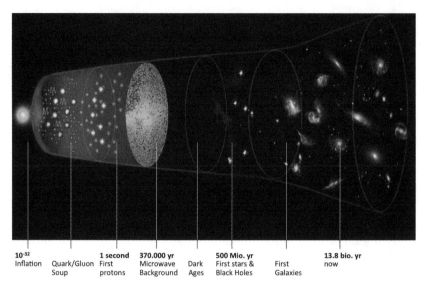

10^{-32}		1 second	370.000 yr		500 Mio. yr		13.8 bio. yr
Inflation	Quark/Gluon	First	Microwave	Dark	First stars &	First	now
	Soup	protons	Background	Ages	Black Holes	Galaxies	

Figure 2.3. The entire history of the universe at a glance. The radius of the visible universe today is shown on an extremely compressed logarithmic time scale. The bottom of the figure shows important events in the history of the universe.

Planck, Heisenberg, and Schrödinger and Einstein's theory of general relativity no longer fit such high-energy densities. The quantum theory of gravity, which is still in its infancy, must therefore describe the state of the universe at this point. Planck units are the characteristic sizes of the embryo universe: Planck time, Planck length, Planck mass, and Planck temperature.[5] At this moment, the observable part of today's universe is minute (see Figure 2.3). As we have already discussed in connection with the Casimir effect, this embryo universe, like any vacuum, contained quantum fluctuations, which will later become extremely important.

After the gravitational force split off, this baby universe began to grow and cool at the same time. If you inflate a bicycle tire, you notice that the air pump and your hands grow warm because gas heats up during compression. Conversely, a ball of gas cools when its volume increases. A dramatic cooling essentially characterized the destiny of the universe, as shown in Figure 2.2. In the beginning, all the particles had energies near the Planck scale that were much higher than the rest masses of every kind of particle known and yet to be discovered, as well as the force-carrying particles. According to Einstein's formula $E=mc^2$, all known and as yet unknown species of particles could therefore continuously form from pure energy and disintegrate again in collisions. When the energy became high enough, for example, two high-energy photons could produce an electron and a positron, the positively charged antiparticle of the electron. Conversely, a particle colliding with its antiparticle would destroy the two matter particles and convert them into radiation energy. In the first fractions of a second after the Big Bang, the density and temperature were so high that all elementary particles, such as photons, neutrinos, electrons, quarks, etc., along with their antiparticles, were continuously converted into each other, establishing a thermal equilibrium between the particles. Therefore, apart from statistical factors, which depend only on the nature of the particles, all particle types initially existed in approximately equal numbers. At this time, the united GUT force, which is supposed to be transmitted by a hypothetical force-carrying particle, the X-boson, dominated. This must have had a very high mass and therefore a very limited lifetime.[6] After a period of about 10^{-34} seconds—an extremely short time but still more than a billion Planck times—the universe cooled down to a temperature of 10^{28} K (kelvin). The X-bosons decayed, and no new GUT-force particles could be generated from the heat bath. This caused a phase transition, a symmetry breaking, in which the strong interaction split off the GUT force and released a large amount of energy. The universe would experience several more such restructuring events while cooling down. The freezing of water in a lake provides a good example of this process. As it grows colder, the originally completely formless water transforms into regularly shaped ice crystals and more complex structures. Such events lead to hypothermia, in which matter reaches lower temperatures than the actual freezing point and experiences an abrupt phase transition, releasing energy. The early universe underwent a similar hypothermia during the inflationary phase.

The GUT-phase symmetry breaking occurred at an energy of about 10^{15} gigaelectronvolts—about four orders of magnitude below the Planck scale but still thirteen orders of magnitude higher than any energy attainable by our largest particle accelerators on Earth. Only the Big Bang itself could have reached such gigantic energies. Much later, about 10^{-10} seconds after the Big Bang, the weak forces and the electromagnetic forces split off from the remaining electroweak interaction. Because modern particle accelerators can achieve this level of energy, this phase transition has been experimentally investigated in detail.

The GUT symmetry breaking is specifically responsible for the excess of matter over antimatter in the universe today, called "baryogenesis." For some unknown reason, there existed a small surplus of about one in a billion quarks over antiquarks and electrons over positrons at the end of the GUT phase, which is responsible for our present normal baryonic matter. How this all works exactly remains unclear. However, the discovery that neutrinos have a finite mass and thus different neutrino families can transform into each other suggests that baryogenesis could have originated in the electroweak interaction. This could explain why the same surplus exists for quarks and electrons so that our universe remains electrically neutral.

The universe has therefore, to some degree, taken a loan in the currency of the weak interaction. However, the symmetry breaking responsible for the baryogenesis leads to the presumption that the protons, which are composed of three quarks and constitute all elementary nuclei, do not live indefinitely but must decay about 10^{35}–10^{36} years after the Big Bang—that is, in the ever-distant future, in order to repay the loan. All the normal matter still present at this time will then melt away like a snowman in the sunlight. Although this will occur almost an infinitely long time from now, the decay of protons should be observable even today if a sufficient number are monitored over an extended time. Scientists conducting the Japanese Super-Kamiokande experiment, in which thousands of light-sensitive detectors monitor a huge underground hall filled with fifty thousand tons of water, have spent several years searching for the characteristic signals of proton decay—so far, without success. The earliest GUT theories predicted a lifetime of about 10^{32} years for the proton, but the Kamiokande measurements already require a lifetime of at least 10^{35} years. Modern GUT theories are now assuming a lifetime of 10^{36} years.

A problem with GUT theories is that they predict magnetic monopoles, strange particles that include an isolated magnetic north or south pole. Magnetic monopoles are not prohibited by the theory; on the contrary, they would be most desirable in order to establish the symmetry between electric and magnetic charges. However, despite an intensive search, they have yet to be discovered in nature. Guth, one of the first scientists to work on magnetic monopoles, found that their energy must be too great for anything but the Big Bang to have produced them. He then developed the idea that the seed from which our entire visible universe sprang must have been so small that it only contained a few magnetic monopoles. These would still exist today, but it is highly unlikely that any will ever come close enough to Earth to be discovered. To explain this scarcity, Guth postulated an early inflationary phase, during which the size of the universe multiplied dramatically within a very short time, ejecting all monopoles beyond our horizon.

Inflation theory assumes that the vacuum in the GUT phase immediately after the Big Bang possessed a much higher energy than the universe today. If this energy density, referred to as the "inflaton field," were expressed as an equivalent mass density, it would have the extremely high value of 10^{89} kilograms per cubic meter—about ten million times lower than the density of the Planck time but with a repelling action that would drive the universe apart. All numerical values for the theory of inflation are still subject to large uncertainties, mainly because GUT-theory physics is not yet fully understood. Therefore, we can use the variables defined in this context as only a rough guideline. In Chapter 1, we saw that a universe dominated by repulsive energy must expand exponentially according to de Sitter's solution of Einstein's field equations.

This "false vacuum" of the GUT time must have decayed. The energy released in the first 10^{-32} seconds after the Big Bang resulted in a dramatic inflationary expansion, giving the resulting fireball its initial momentum. We can compare this process to a boat trip over the Niagara Falls. Imagine you are floating comfortably in a boat somewhere upstream of the falls. The river flows sluggishly, the weather is nice, and there is no indication that you are, in reality, two hundred feet above the lower reaches. When you hear the roar of the falls, you realize the dramatic potential energy residing in the sixty meters of cascading water, which is now converted into kinetic energy in a phase transition. In the lower reaches of the Niagara Falls, the river flows as slow and leisurely as the one above, and nothing indicates that it is at a much lower energy level.

We can calculate the size of the universe visible from Earth, known as our horizon, fairly accurately. It corresponds to the distance that light has traveled back to Earth in the 13.8 billion years since the Big Bang times the amount by which the universe has expanded since the light was initially emitted. Overall, the horizon is about three times the size corresponding to the age of the universe, namely about forty billion light-years, or $3.7 \cdot 10^{28}$ cm. During the inflationary GUT phase transition, which lasted about 10^{-34} seconds to 10^{-32} seconds after the Big Bang, the universe corresponding to our present horizon expanded in the shortest possible time by more than thirty orders of magnitude— from a miniscule spot not much larger than the Planck length to about the size of a soccer ball. During this tiny interval, the expansion happened much faster than the speed of light. At first glance, this seems to contradict Einstein's theory of relativity. However, because the space-time expanded together with its entire contents of particles and energy, this does not pose a problem. The theory of relativity forbids only relative velocities between matter and space that are larger than the speed of light. Following Swiss astrophysicist and cosmologist Gustav Tammann's example of a loaf of yeast bread with raisins that rises in the oven, the whole loaf can, in principle, expand at any speed as long as the raisins move no faster than the speed of light relative to the dough. The energy from the GUT phase transition reheated this "dough" to about 10^{28} K after a temporary hypothermia (see Figure 2.2), producing all the particles from which the hot universe evolved. One can imagine that inflation dramatically tore apart the previous vacuum quantum fluctuations. The pairs of virtual particles comprising the fluctuations became separated during this process and were unable to find each other to annihilate again. The virtual particles of the vacuum, which we have already discussed in the context of the Casimir effect, were therefore converted into real particles. We will later learn about a similar process in the context of Hawking radiation from black holes (see Chapter 8). In this way, the process of inflation froze the quantum fluctuations existing just before the GUT phase transition and spatially expanded them into the primordial fluctuations of the fireball, from which formed the large-scale structures, filaments, clusters, galaxies, and ultimately, stars and planets, as described in Chapters 4 and 5.

Although the theory of inflation was originally developed in the context of elementary particle physics, it has immediately led to the solution of two fundamental problems of Big Bang cosmology known as the horizon problem, discussed in detail in Chapter 4, and the flatness problem. At the beginning of this chapter, we saw that according to recent measurements, the total mass-energy density of the universe—the sum of dark matter and dark energy—is within 1 percent of its critical density so that Ω is almost equal to 1 and therefore, the universe is flat. If the energy density was exactly critical at the time of the Big Bang, then it remained exactly at $\Omega = 1$ throughout the whole expansion of the cosmos. If, however, Ω had been a tiny bit different from 1 at the time of the Big Bang, our present universe would not exist. If the overall density had been a tiny bit larger than the critical density shortly after the Big Bang, our universe would have collapsed in a split second. If, on the other hand, the density had been slightly smaller than critical, the universe would have expanded so rapidly that no galaxies could have formed. Thirty years ago, it was already known that Ω today must have a value between 0.1 and 2 and thus is relatively close to 1. Extrapolating this information back to a time about one second after the Big Bang, it follows that Ω would only have been allowed to deviate from 1 by the tiny fraction of 10^{-15}. Thus, this number belongs among the most accurate quantities ever measured in physics. The original cosmological model, however, contained nothing to explain such an exact fine-tuning of a flat universe.

Inflation, on the other hand, has no problem whatsoever doing exactly this: imagine a small, shriveled balloon, which obviously has a rough, curved surface. When you inflate this balloon, all its wrinkles smooth out as it stretches. If you blow up this balloon until it is as big as Earth, then you will not be able to see the slightest curvature on its surface—it has become absolutely flat for all human recognition. Because inflation blows up the tiny primordial universe by a gigantic factor, it simultaneously makes it absolutely flat. For decades, theoretical cosmologists always assumed $\Omega = 1$, while observational cosmologists were unable to scrape together enough matter in the cosmos to come up to the critical density. The recent determination of the total energy density in the universe, which deviates by less than 1 percent from the critical density after adding up dark matter and dark energy, suddenly provides an excellent confirmation of the inflation theory and has added to its credibility. It rarely happens that bizarre theories make predictions that are confirmed as accurately years later.

According to the theory of inflation, the horizon of our visible universe can by no means represent the whole structure. This can be illustrated by imagining you are looking at the horizon while on a sandy white beach in Hawai'i. Although in former times people believed the horizon to be the true end of the world and thought they would fall off if they traveled there, we now understand that the horizon is an optical illusion. If we take a boat to the horizon, we find yet another horizon. Indeed, one can conclude from the flatness of the horizon that Earth must be huge. The horizon is only about ten kilometers away and therefore represents a tiny fraction of Earth's entire surface. If you lived on the asteroid B612 like the little prince from Saint-Exupéry's story, your horizon would be highly curved, and you would know that you lived on a fairly small planet. If you apply this analogy to our visible universe, which is also limited by a very flat horizon due to the finite travel time of light rays since the Big Bang (13.8 billion years ago), you conclude that it represents just a tiny fraction of a much vaster structure bloated up by inflation. If you compare the tip of a needle to the size of Earth, you are still off by many orders of magnitude. A better match would be the needle tip in relation to the distances between the galaxies. We get a small sense of the universe's gigantic size from Figure 1.3. Inflation thus answers the popular and notorious question of what lies outside our visible universe: more universe. Douglas Adams' *The Restaurant at the End of the Universe* is therefore as much a metaphor as the pot of gold at the end of the rainbow.

Starobinsky's school in Moscow also educated Andrei Linde, who has developed inflationary theory furthest in recent decades. Linde, who now works at Stanford University, coined the idea of "chaotic inflation," which states that our universe is just one of many entities in a so-called multiverse. According to Linde, new inflationary universes continuously arise from the chaos of quantum fluctuations. This could happen any time, even within our own universe and while you read this book. Perhaps the multiverse can be imagined as a sea of ballooning bubbles, from which in turn swell bubbles over bubbles, similar to the foam of a bubble bath. Different physical laws govern each of these universes. Many may be "too heavy" and collapse right away. Others might immediately explode due to too much energy. But some of these universes would possess balanced physical boundary conditions that allow the formation of matter, galaxies, stars, planets, and ultimately, intelligent species who can think about the wonders they live in.

In a lecture celebrating the opening of the Arnold Sommerfeld Center at the Ludwig-Maximilians-University of Munich in 2005, Linde compared the multiverse to a rugged mountain landscape with many steep pinnacles and a few flat pastures and valleys in between. This is the string landscape model by Leonard Susskind from Berkeley. Looking at this landscape from above, one is not surprised that green grass, animals, and people live only on the flat slopes and in the valleys—the laws of physics do not allow them to permanently reside on the sheer cliffs. Likewise, our place in the universe could be due to a selection effect: complex structures and people exist only in areas with favorable physical boundary conditions. With this vision, which is also referred to as the "weak anthropic principle,"[7] Linde prepares us for the next Copernican revolution. Not only was man exiled from the center of the solar system by Nicolaus Copernicus and from the center of the Milky Way by Harlow Shapley; not only does the substance from which we are all made account for just 6 percent of the energy-matter-density and we remain clueless about the remaining 94 percent: it could be that our vast universe represents only the tiniest fraction of a mountain-, forest-, and meadow-"verse" that exists in a giant web of interpenetrating "verse" bubbles.

Before you now slam this book shut, get frustrated, and turn your back on physics once and for all, I would like to call out to you, "Hold on! After all, we are the only intelligent species in the known (to us) universe, and after all, the stuff we are all made of is so unique that it's worth thinking about! If not us, then who? If not now, when?" So I hope that you will remain with me throughout this "limitless journey."

THE QUARK SOUP

After the inflationary period, the baby universe essentially consisted of an ever-expanding and cooling dense soup of elementary particles. Free quarks and gluons, as well as leptons, such as electrons, neutrinos, photons, and unknown dark matter particles floated in this soup. All particles with a rest mass smaller than the ever-decreasing thermal energy scale were continuously converted into each other. This was, of course, also true for the respective antiparticles. Such a state of matter is referred to as a "quark-gluon plasma." About one-millionth of a second after the Big Bang, the quark soup cooled from its original temperature of 10^{28} K down to 10^{14} K. As the quark-gluon soup grew colder, the annihilation of the quarks and the antiquarks picked up speed, outweighing the corresponding formation processes, and a dramatic quark-slaughtering process ensued. At the end of this process, almost every quark found a corresponding antiquark and vanished with its partner into two photons or an electron-positron pair. Had it not been for the tiny symmetry breaking in the GUT phase—the loan that the universe took out at the very beginning—all quarks and antiquarks would have dissolved into thin air and I would not be sitting here writing a book. But thank God, one out of every billion quarks failed to find a counterpart for its destruction and therefore survived the carnage.

By the end of the quark era, about one millisecond after the Big Bang, the universe cooled down so far that a phase transition from a quark-gluon plasma to the hadron plasma took place. All particles that consist of quarks, particularly the baryons, such as protons and neutrons, and the volatile mesons, such as π-mesons, are called "hadrons."

At an energy scale of about one gigaelectronvolt, which corresponded to the mass of protons and neutrons, the energy of the background radiation could no longer sufficiently overcome the attractive forces of the gluons. In each case, three quarks combined to form protons and neutrons, the fundamental building blocks of atomic nuclei usually learned about in high school. This way, the first complex structures in the universe formed. Later, after the universe cooled down further, hydrogen atoms formed from a proton and an electron and helium atoms formed from two protons, neutrons, and electrons. We also know from school that almost the entire mass of an atom is gathered inside the atomic nucleus. Expressed in terms of energy, for example, the proton has a mass of 938 MeV/c^2 (megaelectronvolt) while the electron, with a mass of 0.5 MeV/c^2, is about two thousand times lighter. A fascinating aspect of the capture of the three quarks within a proton is the fact that the three quarks together have a much lower rest mass than the proton. The up-quark with 4 MeV/c^2 and the down-quark with 8 MeV/c^2 together weigh only a small fraction of what the proton weighs. Most of the proton and neutron mass and therefore, our normal baryonic matter, must be stuck inside the baryons in the form of kinetic energy from the quarks and binding energy from the gluons. In fact, there is high life within the proton! The quarks race back and forth, held together by gluons. In addition to the three normal quarks, many virtual quarks and antiquarks are constantly created and destroyed again—a coming and going such as that found at a train station. Overall, the proton stores a lot of energy, which betrays itself to the outside world as gravitational mass. Accordingly, the mass of all normal matter can be interpreted as captured Big Bang energy. This really is applied special relativity following Einstein's famous equation $E=mc^2$!

NEUTRINOS, DARK MATTER, AND ELECTRON-POSITRON PAIRS

The early universe at this point was essentially made up of protons and neutrons, electrons and positrons, photons and neutrinos, and the dark matter particles still to be discovered. Meanwhile, all four known fundamental forces had separated from the GUT force, but the matter was still so dense that not even the neutrinos could escape. If you try to shield neutrinos with normal matter, you need a wall made out of lead at least one light-year thick. You can imagine how dense the universe must have been a millisecond after the Big Bang. At this point, the visible part of what would become our present universe was about as large as our solar system and dense enough to block neutrinos.

A short time later, the density of the universe dropped low enough to finally release the neutrino ghost-particles. Today, nothing blocks their way, and they continue to fly in their last chosen direction at almost the speed of light. Every second, about ten million neutrinos from the Big Bang penetrate our bodies. They can easily pass through even the whole Earth. Very rarely, a high-energy neutrino comes so close to an atomic nucleus that it weakly reacts and strikes out a very energetic particle from the nucleus. In water or in very transparent ice, this particle produces a blue flash of light, called Cherenkov radiation, which can be traced with sensitive photon detectors. It is difficult, however, to distinguish these events from those arising from many other high-energy particles, which nature can produce using a wide variety of processes. In the Antarctic ice and in one of the clearest areas of the Mediterranean Sea, gigantic neutrino detectors with such evocative names as AMANDA (Antarctic Muon and Neutrino Detector Array), ANTARES (Astronomy with a Neutrino Telescope and Abyss Environmental Research Project), and IceCube detect Cherenkov radiation using huge amounts of seawater or gigantic volumes of the purest ice near the South Pole. Except for the neutrinos from supernova 1987A in the Large Magellanic Cloud, as well as the regularly measured neutrinos from the sun, no cosmic neutrino has yet been detected. Due to the expansion of the universe, the energy of the neutrinos from the Big Bang must have dropped to a temperature of about 1.95 K above absolute zero.[8] The detectors used so far are not sensitive enough for such low-energy neutrinos. Should it ever be possible to take a picture of the cosmic neutrino background, one could directly observe the state of the universe about a tenth of a second after the Big Bang.

As described in Chapter 1, we unfortunately still do not know anything about the dark matter particles. Most likely, however, these are massive particles that follow only the weak interaction, just like the neutrinos. If this is the case, the dark matter decoupled from the rest of the universe around the same time as the neutrinos. In contrast to neutrinos, which are very light and move nearly at the speed of light, dark matter particles must be much heavier and move relatively slowly.

We therefore speak of "cold dark matter." Such massive particles often collided within the dense, hot soup of the early universe. We assume that dark matter exists in equal numbers of particles and antiparticles, similar to all other types of matter formed during the Big Bang. When two of these particles get close enough for a weak interaction to occur, they can annihilate each other. Particles and antiparticles of dark matter, which originally must have been as numerous as photons, should already have mostly destroyed each other by the time of the neutrino decoupling. If the dark matter particle is actually a weakly interacting massive particle (WIMP) with a rest mass ten to one hundred times larger than a proton, only a small fraction should have remained by that time. But obviously, enough would still persist to dominate the matter balance today. As we will see later, dark matter began to form the first large-scale structures in the universe immediately after its decoupling under the influence of its own gravity.

More and more lightweight particles spontaneously formed from the high-energy photon bath before being destroyed again, especially electron-positron pairs. The positron is the positively charged antiparticle of the electron. Both have a rest mass of 511 keV/c^2 (kiloelectron-volt). If an electron and a positron near each other, they are attracted through the electromagnetic force, dance around each other briefly, and then annihilate, emitting two photons of energy 511 keV. At about one tenth of a second after the Big Bang, the temperature of the universe dropped below the energy scale of the electron-positron pair: 1 MeV/c^2, or about 10 billion K. The photon bath could no longer generate any new pairs, and a dramatic destruction process began. Due to the electron-positron annihilation, the cosmos heated up and would, in the future, always be about 40 percent warmer than the neutrinos, which had already decoupled. Almost all electrons and positrons annihilated each other in the same way as the quarks, and about one in one billion electrons remained. As if by sheer magic, and so far still completely unexplained by the standard model of particle physics, exactly as many negatively charged electrons remained as needed to balance the remaining protons' positive electrical charge. As it turns out, the universe is electrically neutral with astonishing precision: because the electromagnetic force is much stronger than the gravitational force but has the same extremely large range, each electric charge excess in the universe would create forces much stronger than the gravitational force, but these are not observed.

THE PRIMORDIAL ELEMENT SYNTHESIS

Because in the beginning nothing was there, everything we see in the universe today must have been formed after the Big Bang. Above all, this includes the chemical elements of the periodic table, which form the basis for all matter in our daily environment. The nuclei of the chemical elements are composed of protons and neutrons. In the neutral state at low temperatures, the surrounding electron shell contains exactly as many negatively charged electrons as the core contains positively charged protons. About one second after the Big Bang, the temperature of the fireball dropped so far that it became comparable to the heat of the nuclear fusion ovens that exist inside stars. As a result, complicated composite atomic nuclei could form without being torn apart again by the energy of the photons. The cosmic abundance of the lightest elements—hydrogen, helium, lithium, and beryllium—yields one of the most convincing confirmations of the Big Bang model. Chemists can analyze the different elements on Earth very well. However, these elements' proportions are not representative of the entire cosmos because Earth's gravity cannot hold the lightest elements; they have already evaporated into space. The second-lightest atom in the periodic table hardly exists on Earth and has only been discovered in the spectrum of the sun. That's why they gave it the name "helium"—after Helios, the Greek sun god. Earth contains enough underground helium reserves to supply large quantities of this element to use in technical applications such as a coolant or as a lifting gas for balloons and blimps. You can perform a nice and rather harmless experiment with helium by inhaling the gas from a balloon and then for a short time talking like Mickey Mouse. However, in recent years, helium has become scarcer and more and more expensive!

US physicist George Gamow, who emigrated from Russia in 1933 under dramatic circumstances,[9] is often considered the father of the Big Bang. We have already come to know the affable, literary Gamow in the context of Einstein's "greatest blunder." In addition to his great scientific achievements, he was a gifted science communicator. He wrote popular science classics such as *One, Two, Three . . . Infinity* and the *Mr. Tomkins* book series but also left an unfinished autobiography, *My World Line*,[10] that is the source of the quote from Einstein. Gamow was one of the scientists of the Manhattan Project, who built the atomic bombs that devastated the Japanese cities of Hiroshima and Nagasaki during World War II. Along with Edward Teller, he laid the first foundations for the hydrogen bomb. After the Second World War, Gamow grew concerned with the origin of the elements in the cosmos. Manhattan Project scientists had analyzed the radioactive fallout of the early nuclear tests and found that the explosions had created entirely new elements and isotopes. Gamow became convinced that all chemical elements had been formed in a similar way in a huge explosion in the early universe. His model of a hot Big Bang thus directly continued the theme of the primordial atom postulated in 1927 by Belgian priest Georges-Henri Lemaître. Gamow was also the first to point out that the extremely high temperatures in the early universe must have left large amounts of radiation behind that should still be visible in space today—but more on that later.

Future Nobel Laureate Hans Bethe was head of the theory department in the Manhattan Project. As early as 1938, he had shown that the fusion of hydrogen with helium essentially produced the energy in stars. He did this simultaneously with but independently of German physicist Carl Friedrich von Weizsäcker. Stars with a mass larger than the sun use the so-called Bethe-Weizsäcker cycle, which uses the element carbon as a catalyst. In contrast to Gamow's assumption, some portion of helium and probably also other elements must have been generated within the stars themselves. Life on our planet has owed its existence to the energy created by nuclear fusion for billions of years. Fusion researchers have been trying for decades to tame this energy and are making progress. Bethe later became one of the most ardent advocates of the international ban on nuclear tests. I had the good fortune to meet him personally in 1992 during my stay as a visiting astronomer at the California Institute of Technology (Caltech), where he was also a guest. He invited me to his office and asked me about the latest results of our X-ray roentgen satellite (ROSAT; see Chapter 8). He was already eighty-six years old but still keenly interested in all the new discoveries in astrophysics. In the spring of 2005, he died at his home in Ithaca at the almost biblical age of ninety-nine. As Gamow published his theory of the origin of the elements in 1948 together with his doctoral student Ralph Alpher, he noticed that their names read almost like the Greek letters alpha and gamma. For symmetry, however, he still lacked the letter beta. Quickly, Gamow added the name of his colleague Bethe to the author list without asking him.[11] This publication became known as the αβγ paper.[12] Bethe, in his modest way, did not complain about the move but gamely took part in further elaborating the theory.

Gamow's strongest opponent was English astrophysicist Sir Fred Hoyle, Eddington's successor on the Plumian chair in Cambridge and one of the biggest critics of the Big Bang theory. Hoyle actually coined the name "Big Bang" as a joking criticism of Lemaître and Gamow's cosmological model. Hoyle and other colleagues such as Herman Bondi and Thomas Gold represented the "steady state" theory, an interesting alternative to the Big Bang model that postulates that the universe has always been as we see it today and always will remain the same. They accepted that the galaxies are moving apart but assumed that new matter is constantly being created between the galaxies. This would require that no more than a few protons per cubic meter be created each year—a quantity so tiny that the possible formation processes remain hidden. Hoyle, who had also worked on the theory of nuclear fusion within stars and in particular on supernova explosions, believed that the stars' energy created all the elements in the cosmos. He knew the relative abundance of the chemical elements on Earth and in the stars and had discovered that the heavier elements are, the less frequently they occur. Only one group of elements with masses near iron clearly stands out from this trend. Hoyle attributed this to iron having the highest binding energy. All elements heavier than iron generate energy using nuclear fission while all lighter elements produce energy using fusion. Regardless of which side you come from, iron is always the final product. This finding served as proof enough to Hoyle that the elements are created in the interiors of stars.

However, both the representatives of Big Bang nucleosynthesis and the proponents of fusion within the stars had a big problem—namely, to produce elements with mass numbers greater than five. Bethe and Weizsäcker had already described the first step, in which two protons and two neutrons unite to form helium. In the second step, two helium atoms would have to combine to form beryllium-8, a nucleus with four protons and four neutrons. But beryllium-8 is unstable and decays immediately into its components. All other nuclei with mass numbers five through eight are unstable as well. Without a stable nucleus, it is impossible to build heavier elements. In his important work of 1939, Bethe therefore concluded that the interiors of stars cannot produce elements heavier than helium. However, the young theorist Ed Salpeter, who worked with Bethe, suggested a solution to the problem in 1952. Just before a red giant star explodes as a supernova, temperatures and densities must prevail within its interior to occasionally allow three helium nuclei to almost simultaneously collide just to create a nucleus of carbon-12. This nucleus is stable and allows the fusion chain to continue and form heavier elements.

For Hoyle, however, this was far from enough explanation. Hoyle had calculated that the probability of three helium nuclei simultaneously colliding in the interior of a massive star was far too low to explain the abundance of the element carbon. This event would be the nuclear physics equivalent of three fully loaded shopping carts crashing into each other in the parking lot of a supermarket: interesting, but extremely unlikely! The element carbon is responsible for Earth's overall biology and thus, life. Without the proper amount of carbon we could not exist. Hoyle therefore argued using the anthropic principle: "I exist, therefore the nucleus of the carbon-12 atom must have a resonance at an energy of 7.7 MeV!" "Resonance" is a trick built into nuclear physics to allow an easier formation of carbon-12. Hoyle found it much more likely that a beryllium-8 nucleus would collide with a helium nucleus. The two nuclei would have to combine to a carbon nucleus before the short-lived beryllium nucleus decayed again. Hoyle's calculation of the resonance energy corresponded exactly to the rest mass of one helium nucleus and one beryllium-8 nucleus, along with their kinetic energy, at a temperature of 100 million K, such as prevails in the interior of a red giant. Such a resonance would cause the beryllium-8 nucleus to connect very easily with a helium nucleus, dramatically increasing the abundance of carbon.

In 1953, Hoyle visited Willy Fowler's group at the Kellogg Radiation Laboratory at Caltech and told them of his prediction.[13] The nuclear physicists at Caltech were initially very skeptical about an astrophysicist telling them anything about the energy levels of carbon. However, Fowler and his colleagues grew enthusiastic after they actually found the resonance line of carbon that Hoyle had predicted at their accelerator. The line was found at an energy of 7.65 MeV, which was less than 1 percent off from the predicted level!

This was the first theoretical prediction based on the anthropic principle, which an experimental verification later confirmed. The strange chance that carbon's resonance achieves exactly the right level in order to be abundant enough is reinforced by the fact that oxygen, the next element after carbon, does not have a resonance at a similar energy. Otherwise, most of the carbon nuclei would have immediately turned into oxygen nuclei. The two most important elements necessary for life, carbon and oxygen, just happen to have properties that allow the formation of more complex structures. This has convinced some astrophysicists that an intelligent plan may be behind the laws of physics. Along with Fowler and the couple Margaret and Geoffrey Burbidge, Hoyle worked at Caltech for a long time on cosmic nucleosynthesis in supernova explosions. The four jointly published a paper in 1957, known by astronomers as "B^{2}FH"—the initials of the authors—that is still a bible for astrophysicists. Fowler received the Nobel Prize for his work in 1983. Hoyle, unfortunately, missed out.

When Gamow and his colleagues realized that no elements heavier than helium could be produced in the Big Bang, they grew frustrated and gave up their research. In his inimitable, poetic way, Gamow admitted his defeat in this regard and published his own version of the story of the Creation in his autobiography:

> In the beginning God created radiation and ylem [primordial matter from the Big Bang]. And ylem was without shape or number, and the nucleons were rushing madly over the face of the deep. And God said: "Let there be mass two." And there was mass two. And God saw deuterium, and it was good. . . . In the excitement of counting, He missed calling for mass five and so, naturally, no heavier elements could have been formed. God was very much disappointed, and wanted first to contract the Universe again, and to start all over from the beginning. But it would be much too simple. Thus, being almighty, God decided to correct His mistake in a most impossible way. And God said: "Let there be Hoyle." And there was Hoyle. And God looked at Hoyle . . . and told him to make heavy elements in any way he pleased. And Hoyle decided to make heavy elements in stars, and to spread them around by supernovae explosions. But in doing so he had to obtain the same abundance curve which would have resulted from nucleosynthesis in ylem, if God would not have forgotten to call for mass five. And so, with the help of God, Hoyle made heavy elements in this way, but it was so complicated that nowadays neither Hoyle, nor God, nor anybody else can figure out exactly how it was done. Amen.[14]

However, Hoyle had originally also reckoned the statement about helium without his host. In 1964, he wrote a paper with Roger Taylor from Cambridge that spelled out very clearly that the abundance of the light elements is not compatible with an origin within the stars. The conversion of hydrogen into helium produces a large amount of energy that must be radiated in the form of light. If all the helium in the cosmos had been formed in the stars, the Milky Way should be about ten times brighter. They suggested that the helium originated in a "radiation universe" similar to Gamow's Big Bang model.

Ultimately, both Gamow and Hoyle were correct. The Big Bang produced the light elements hydrogen; deuterium, a modification of hydrogen from one neutron, one proton, and one electron; radioactive tritium, a nucleus of one proton and two neutrons; and helium, as well as the rare elements beryllium-7 and lithium-7. The centers of stars and supernova explosions generated all of the heavier elements. In his fascinating book *The First Three Minutes*,[15] Nobel Laureate Steven Weinberg describes in detail the reactions that led to primordial nucleosynthesis. From the relative abundance of the light elements, scientists can very precisely determine the conditions during the first three minutes after the Big Bang. Nucleosynthesis thus represents a central pillar of the Big Bang model.

But back to our fireball: the most important factor determining the fate of the universe was the ratio between neutrons and protons, which in turn depended on the ratio between baryons and photons. We recall that the proton consists of two u- and one d-quark, while the neutron is made up of one u- and two d-quarks. Due to thermal equilibrium, almost exactly the same number of neutrons and protons must have remained after the quark-gluon plasma converted to the hadron plasma. The neutron's rest mass is slightly higher at 939.6 MeV/c^2 than the proton at 938.3 MeV/c^2. This small difference produces a very large effect: although a proton can live almost infinitely, a free neutron has a lifetime of only about eleven minutes, after which it decays into a proton, an electron, and an antineutrino. The universe at this time, however, was only 0.1 seconds old and thus too young for neutron decay. In the thermal equilibrium of the fireball, protons and neutrons could convert into each other with the help of electrons and neutrinos. According to the laws of thermal equilibrium, these conversion processes depend on the energy of the particles involved. Because of the mass difference, it is therefore slightly easier to transform a neutron into a proton than vice versa. However, the total energy of each particle, its rest mass together with its kinetic energy caused by the temperature, must be considered. As long as the thermal energy dominates, the small difference in the rest masses is irrelevant. But, when the temperature falls to values close to the difference in mass between the neutron and the proton, such as 1.3 MeV/c^2, this mass difference becomes very important. The ratio between neutrons and protons therefore depends exponentially on the temperature.

As it grows colder, neutrons decrease in number compared to protons. One second after the Big Bang, at a temperature of roughly 10 billion K, about 20 percent of the neutrons and 80 percent of the protons were left. At this time the neutron's limited lifetime still had not played a role. The universe had to hurry up and form heavier elements by baking protons and neutrons together before the neutrons decayed. As it turned out, this would be a race against time.

Nuclear fusion occurs when the density and the temperature of a plasma is so high that two nuclear particles collide and meld together due to a strong interaction with a range of only about one femtometer (10^{-15} meters). At the same time, the temperature must be low enough that photons from the background radiation do not break the particles' bond. The temperature in our fireball has now dropped so low that the first new nuclei can be made from one proton and one neutron. This hydrogen isotope (a kind of "brother" of hydrogen) is called "deuterium." Much later on Earth, deuterium will be the basis for the heavy water used as a coolant in nuclear fission reactors and as a fuel in fusion reactors. As the next step in the fusion chain, the volatile helium-3 might have formed in the collision of two deuterium nuclei by attaching another proton to one of them. This would have become the first new element after hydrogen! However, if another neutron attached to deuterium, the unstable hydrogen isotope tritium would have formed: a nucleus with a proton and two neutrons. Another collision with a deuterium nucleus could then finally form the very stable nucleus of helium-4. The temperature then fell so low that helium-4 as well as the weaker-bound nuclei tritium and helium-3 survived. Deuterium's weak bond posed a problem, however, because the photons in the background radiation immediately tore it apart. It is about nine times easier to split up a deuterium nucleus than it is to remove a single particle from the nucleus of helium-4. Therefore, heavier nuclei still failed to form. And the neutrons continued to run out of time!

As soon as the temperature dropped to less than 0.9 billion K, which corresponds to the binding energy of deuterium, everything happened very rapidly. The universe at this time was three minutes and forty seconds old (which led to Weinberg's book title). The ratio between neutrons and protons reached a value of about 12 to 88 percent. Deuterium became stable and built the nuclei of tritium, helium-3, and helium-4 in collisions. More sophisticated processes produced the elements lithium and beryllium-7. Although just a bit earlier they would

have been doomed to decay (after eleven minutes), all remaining neutrons were very quickly incorporated into helium nuclei, where, with
a few exceptions, they are still in good hands today. Fortunately, those
bound neutrons remained stable because as fermions they had to obey
the exclusion principle established by Wolfgang Pauli, which states that
every cell in phase space can only be occupied by one fermion. The
densely packed atomic nucleus contains basically no "space," or phase
space, left for the neutrons to disintegrate. (As we will learn later, the
same mechanism stabilizes whole stars made up of neutrons—the so-
called neutron stars.) The fact that deuterium's binding energy was just
high enough to save all the neutrons needed for the heavier elements
is another one of the coincidences that makes our existence possible.
If this binding energy had been just a little bit lower, it would have
taken more than eleven minutes for the universe to cool down enough,
and almost all neutrons would have decayed before the primordial
fusion could occur. In that event, no heavier atomic nuclei and of
course, no people, would exist.

From optical and radio spectra of stars and interstellar gas clouds,
the primordial fraction of helium in today's universe could be determined to be 23.5 ± 0.5 percent, based on the total mass of hydrogen
and helium atoms. Because each helium atom contains two neutrons,
one can accurately determine the ratio between neutrons and protons
during the primordial nucleosynthesis at a temperature of 0.9 billion K.
The fraction of neutrons must have been 12.75 ± 0.25 percent. Because of the thermal equilibrium, the neutron fraction depended exponentially on the temperature and hence directly on the ratio of the
number of baryons (neutrons plus protons) to photons. Similar rela-
tionships exist for the deuterium still remaining from the Big Bang,
as well as the isotope helium-3 and the elements lithium and beryllium. Because we know the precise number of photons in the universe,
as Chapter 3 will discuss, we can determine the density of baryons in
the universe directly from the abundance of the light elements. We recall that from about one billion quarks originally existing in the primordial soup, only one survived the large quark annihilation after the
Big Bang. Because the fireball must have contained about as many
quarks as photons during the thermal equilibrium, and these photons
still exist, the ratio between baryons and photons today must also be
approximately one to one billion. From this ratio, the mass density of
baryons can be derived from the light element abundance. Expressed
in the units of critical density defined above, a value of $\Omega_{baryon} = 0.06$

is found. That is, 6 percent of the mass-energy density of the universe consists of the stuff we are made of.

This is the basis for the third Copernican revolution! The nucleosynthesis of the light elements not only yields an excellent confirmation of Gamow's hot Big Bang model but also demonstrates the fact that 94 percent of the mass and energy density of the universe must be made from other substances than we are made of. The baryons of our daily life comprise only a small part (one-sixth) of all the matter in the universe, which adds up to an energy density of $\Omega_m = 0.30$, as we have seen. About 25 percent of the universe's total energy density consists of the unknown dark matter. Table 1 in the Appendix and Figure 1.8 summarize the currently known distribution of the universe's mass-energy density.

Chapter Three

CLEARING UP

THE CAPTURE OF THE ELECTRONS

After the primordial nucleosynthesis, the universe consists mainly of protons and helium nuclei in a mass ratio of 77.7 to 22.3 percent, as well as the exact number of free electrons necessary to keep the fireball plasma electrically neutral. Dark matter particles, together with the neutrinos, have already decoupled from the rest of the universe and now live a life of their own. Numerically, however, the neutrinos and the photons dominate. As we saw in Chapter 2, there are about a billion times more neutrinos and photons than baryons and electrons. The photons continue to play the leading role, which is why this state of the universe is referred to as "radiation-dominated." Despite their relatively small number, the electrons make life difficult for the photons. Their scattering cross section—the area that blocks the photons—is so large that light cannot pass through this hot plasma. Again and again, the electrons scatter the light rays, leaving the cosmos as opaque as a dense fog. Because of the short mean free path available to the photons between scatterings, they couple strongly to the electrons, which in turn firmly bond to the protons and the helium nuclei through electromagnetic attraction. The photons and the baryons therefore form a kind of joint liquid, or photon-baryon fluid.

The cosmos remains in this state for a very long time—about 380,000 years—during which it continues to expand and cool. At some point in time, when the young universe was about 1,100 times smaller than it is today, it cooled down to 3,000 K, a temperature comparable to the surface of the sun (5,500 K) or the inside of a flame. The hottest-known flame, with a temperature of about 4,800 K, is obtained when the highly toxic carbon-nitrogen compound cyanogen $(CN)_2$ is burned with pure oxygen. The flame of a gas stove has a temperature of roughly 2,300 K, whereas a candle flame measures only about 1,200 K.[1] At a temperature of 3,000 K, the universe passes through another phase transition, a kind of freezing process, that dramatically changes its appearance. Well above this temperature, which corresponds to an energy of about 0.25 electronvolts (eV), stable hydrogen atoms can form because their binding energy is 13.6 electronvolts. However, because the ratio of photons to electrons is about one to one billion, the atomic nuclei and electrons remain separated down to considerably lower temperatures. As we already know, this is a plasma state. As soon as the temperature drops below 3,000 K, protons and electrons combine to form hydrogen atoms, and a little later, helium nuclei and electrons together form helium atoms. The scattering cross section of the hydrogen and helium atoms is much smaller than that of the free electrons. Abruptly, the process of forming atoms clears the way for the photons, which can now move freely through

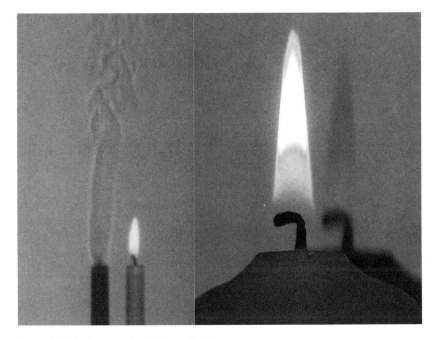

Figure 3.1. Shadows produced by candle flames.

space—and the fog clears and the universe becomes transparent. From now on, the matter and the photon bath will develop independently, without significant interaction. The universe has left the radiation-dominated era and has now entered the matter cosmos era.

All of this sounds highly complicated and physical. We can, however, observe this phase transition rather easily using a simple candle flame. A candle flame is not a particularly good model of the early universe fireball, especially because it is lower-temperature dusty plasma contaminated with soot particles. Nevertheless, the crucial phase transition between the plasma and the neutral gas is very similar. Figure 3.1 (left) shows the shadow of a candle flame. The image clearly demonstrates that hot, burning gas streams upward from the shining flame so that even far above the fire, you can burn your fingers or ignite a match. The right image in Figure 3.1 zooms into the flame. Amazingly, the flame casts a shadow; it is partly opaque. The sharp boundary of the candle flame marks the transition from the opaque plasma to the transparent gas state.

THE HUNT FOR THE BACKGROUND

Clearing up in the early universe is an almost everyday occurrence. In this phase, protons and nuclei combine with electrons to form atoms, a process referred to as "recombination," based on similar processes in laboratory plasmas. However, the prefix "re-" is misleading, as this event, after all, describes the very first combination of elementary particles into atoms. Ever since the era of decoupling, also called the "epoch of last scattering," all photons in the universe move in the direction toward which they were last scattered. Even today, these photons should travel freely through space. But where are they? After all, the sky is pitch black at night! This is the basis for the famous Olbers' paradox—the solution of which, interestingly enough, Edgar Allan Poe anticipated first.[2] The trick lies in the expansion of the universe. Since the time of decoupling, the expanding universe has reduced the energy of the photons by a factor of roughly one thousand due to dilution. During that time, the initial temperature of 3,000 K must have dropped to only a few degrees kelvin above absolute zero (see Figure 2.2). If we look into the depths of the cosmos and thus back to the beginning of time, we can see directly within the hot plasma ball and observe its state. We see a wall of hot plasma, similar to the sharp boundary of the candle flame (see Figure 3.1), where light was scattered for the last time. It moves away from us at many times the speed of light; therefore, the light it has emitted has strongly redshifted. We see that the original temperature of about 3,000 K has "cooled down" to about 3 K. As we discussed in Chapter 2, such superluminal speeds, which Einstein's theory of special relativity explicitly forbids, are actually possible in the theory of general relativity. Einstein's theories only forbid velocities larger than the speed of light relative to the underlying space. The entire space enlarges, however, when we look at the expansion of the cosmos while the observable matter practically stands still. The apparent superluminal speed fits with our observations.

In 1948, George Gamow and his colleagues Ralph Alpher and Robert Herman were the first to recognize that the hot Big Bang, which created the chemical elements, must have emitted photons with the spectral shape of a blackbody that still exist in the universe today. They predicted an "echo" of the Big Bang and estimated its temperature to be between 3 K and 10 K, which is amazingly close to the value of 2.73 K we know today. Alan Guth mentions in his book that a lot of luck was involved in this estimate because the quantities in this calculation—such as the average density of the universe and the details of nucleosynthesis—still had big question marks hanging over them at that time. However, this should not at all diminish Gamow's achievements.[3] Because Gamow's team failed to continue its original line of research due to the difficulties with nucleosynthesis, these predictions were gradually forgotten. As described in Chapter 2, much later in 1964, Fred Hoyle, who actually opposed the Big Bang theory, and his colleague Roger Taylor also assumed that helium must have been formed in a hot radiation cosmos. But they failed to take the final step to calculate the temperature of this radiation. At any rate, none of the pioneers in the field of nucleosynthesis came up with the idea of actively looking for the relic radiation of the Big Bang. On the other hand, in the late 1950s and the early 1960s, there were no detectors sensitive enough for cosmic microwave radiation.

Therefore, more or less by chance, finally proving the Big Bang theory fell into the lap of two radio astronomers, Arno Penzias and Robert Wilson. In a laboratory that the Bell Telephone Company originally built in Crawford Hill, New Jersey, for satellite communications, the two young scientists were tasked with optimizing a very sensitive horn antenna, six meters in diameter, for radio astronomy. The antenna worked in the radio range at a wavelength of 7.35 centimeters. In order to detect very faint radio signals from celestial sources, all other noises from the antenna system itself and the environment had to be very well understood, evaluated, and subtracted. Noise from the radio antenna's own electrical circuits, for example, can be measured and subtracted by directing the antenna alternately between the celestial object and a very cold calibration reference source. For this purpose, Penzias and Wilson used the opening of a thermostat filled with liquid helium at a temperature just 4.2 degrees above absolute zero. Earth's atmosphere provides another source of background radiation, producing a radio signal at a temperature of about 20°C. The two researchers tracked this atmospheric noise by alternately pointing the antenna in different directions between the zenith and the horizon. In spring 1964, Penzias and Wilson were at a point where they had all possible background noise sources under control. They had even removed bird feces from their antenna, assuming they might disturb the signal. But regardless of which portion of the sky they looked at, and regardless of the time of day or the year, they were unable to reduce the signal to zero even after subtracting all noise sources. No matter what they did, a constant, mysterious signal roughly 3.5 degrees higher than expected interfered on their antenna. Because they lacked access to any of the cosmological interpretations, they were reluctant to publish these results before they could explain this signal.

At the same time, and completely independently, a group of scientists at Princeton University led by radar specialist Robert Dicke set out to systematically search for cosmic background radiation. Among other things, Dicke had developed an extremely sensitive microwave radiometer during World War II. Independently of Gamow and Hoyle, he became convinced that relic radiation from the early stages of the universe must still be present. He also based his argument on the chemical element abundance, but his ideas contradicted Gamow's. He believed that the universe's energy density is much higher than its critical density so that it oscillates periodically. In this scenario, the universe's expansion regularly stops and turns into a cataclysmic collapse called the Big Crunch, followed by a new Big Bang, the Big Bounce. Dicke learned from Princeton astrophysicist Martin Schwarzschild—the son of Karl Schwarzschild, whom we will come to know in the context of black holes—that some stars contain hardly any elements heavier then hydrogen and helium. These stars must therefore have formed from almost virgin material. In an oscillating universe, the heavy elements created during previous cycles should still be present after each bounce. From the absence of these heavy elements in certain stars, Dicke concluded that the temperature at each Big Crunch was so high that it disassembled all the chemical elements and forced the chemical evolution in each cycle to begin again virtually from scratch. In hindsight, Dicke's idea shared many similarities with Gamow's, but Dicke motivated his team to systematically search for the radiation from the hot early phase of the universe. He inspired two young experimental physicists, Peter Roll and David Wilkinson, to install a small Dicke radiometer on the roof of a building at Princeton University. With a thirty-centimeter opening, it was a lot smaller than Penzias and Wilson's horn antenna, but it also already had a cold reference source. Dicke asked theoretical physicist Jim Peebles to calculate predictions regarding the observability of the background radiation in today's universe using the model of the Big Bounce. Completely independently from Gamow, Peebles also concluded that this had to involve blackbody radiation. He estimated its temperature to be about 10 K. Early in 1965, Peebles submitted a paper titled "Cosmology, Cosmic Black Body Radiation, and the Cosmic Helium Abundance" to the journal *Physical Review*.

Despite several correspondences with the publisher, however, this paper was never accepted for publication, mainly because an anonymous referee complained that it did not address earlier work on the subject, in particular by Gamow's group. By this time, however, Peebles had already given a number of scientific presentations on his work, and as word spread within the scientific community, a connection was quickly made between the groups at Princeton and in Crawford Hill. During a visit to the Bell Laboratory and the six-meter antenna, Dicke and his colleagues immediately became convinced that Penzias and Wilson had discovered the echo of the Big Bang. They promptly agreed to publish two separate papers in the *Astrophysical Journal,* back to back, that referenced each other. Although Penzias and Wilson were still skeptical regarding the cosmological interpretation of their results and exclusively focused on the technical details of their measurement,[4] Dicke and his colleagues' paper discussed the finding's cosmological relevance.[5] Penzias and Wilson later received the Nobel Prize for this discovery. Gamow (who by this time was deceased), Dicke, Peebles, and their colleagues went away empty-handed. In his book *Light from the Edge of the World,* my teacher Rudolf Kippenhahn related this story with his inimitable dry humor: "One group predicted it, another group looked for it, and a third group knew nothing about the other two and found it."[6]

In retrospect, and upon closer inspection, evidence of the radiation discovered by Penzias and Wilson can be found in previous astrophysics research. In 1941, Andre McKellar found rotational excitations in the interstellar molecules of the cyanide radical CN that pointed to a very low excitation temperature of 2.3 K.[7] French radio astronomer Emile Le Roux probably directly saw the background radiation in 1955 while researching his doctoral thesis. He carried out an all-sky survey at a wavelength of 33 centimeters and derived a temperature of 3 ± 2 K.[8] Ukrainian astronomer Tigran Shmaonov found a temperature of 4 ± 3 K at a wavelength of 3.2 centimeters in 1957. Interestingly, you yourself can hear and see residual radiation from the Big Bang relatively easily on the radio or the television. If you set your radio to a frequency between the stations, you will hear a static noise, and if you turn the television on without tuning to a station, you will see the famous "TV snow." A small fraction of this noise (about 1 percent in the television) comes directly from the Big Bang. In any case, the radio astronomers' discovery was enthusiastically received, and within a short time, measurements were carried out at several other wavelengths compatible with a temperature of about 3 K. Figure 3.2 shows a blackbody spectrum at 2.7 K. But measurements in the vicinity of maximum radiation—particularly in the area of the downturn at higher frequencies and wavelengths in the millimeter range—still proved difficult due to inadequate detector technology and the influence of the atmosphere. Because water molecules strongly absorb microwave radiation—a fact the home microwave oven takes advantage of—measuring microwave background radiation from the ground is a challenge. The first measurements in the vicinity of maximum radiation were indirectly conducted based on the absorption lines of cyanogen in the optical spectrum of stars.[9] Similar to McKellar's measurements, this molecule, which is found in interstellar clouds and thereby is exposed to background radiation, can be used as a kind of thermometer. Measurements above the maximum were previously only possible with high-altitude research balloons and rockets and contained large errors.

In 1974, several independent groups at NASA proposed, as part of the Explorer program, a satellite mission to exactly measure the microwave background radiation in space. In 1976, the decision was made to construct the Cosmic Background Explorer (COBE). Thirty-year-old John Mather of the Goddard Space Flight Center was named project scientist. Hundreds of scientists and technicians worked for approximately ten years to realize this dream. The scientific payload

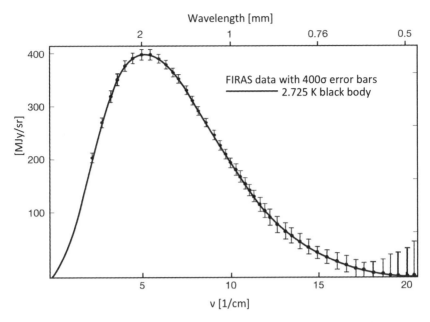

Figure 3.2. The spectrum of the microwave background radiation measured by the Cosmic Background Explorer (COBE). The error bars shown were enlarged by a factor of four hundred to illustrate the extremely high accuracy of the measurement. With utmost precision, the measured spectrum agrees with the theoretical curve formulated by Max Planck for a blackbody radiator, with a temperature of 2.725 ± 0.01 K. (Courtesy of the NASA/ COBE team.)

consisted of three instruments developed by various institutes and universities: the Far-Infrared Absolute Spectrophotometer (FIRAS) was supposed to measure the spectrum of the background radiation with unprecedented precision; the Differential Microwave Radiometer (DMR) was intended to image minute spatial fluctuations in the background radiation using two antennas tilted at fifteen degrees to each other; and the Diffuse Infrared Background Experiment (DIRBE) was a Gregory telescope with a nineteen-centimeter aperture to measure the diffuse infrared background. Originally, NASA planned to launch COBE using a US Delta rocket, but in the course of developing the space shuttle program, NASA decided to launch it with the shuttle. After the long delays and funding shortfalls that commonly occur with extremely complex scientific space missions, COBE was ready for launch in early 1986.

The construction of COBE happened almost in parallel with the development of the roentgen satellite (ROSAT) mission in which I was involved from about 1984. ROSAT was originally scheduled to launch in 1987, also via space shuttle. In January 1986, my former doctoral thesis supervisor Joachim Trümper, the head of the X-ray group at the Max Planck Institute for Extraterrestrial Physics in Garching, Germany, invited some foreign guests and colleagues to his home for the evening on the day the space shuttle *Challenger* exploded. We were all shocked and very saddened as we watched it occur on television. The characteristic forked white explosion cloud has been forever etched in my memory. But we were also very concerned—what would happen to NASA's space program and the planned satellite launches, which were all put on hold? Since the shuttle launch of the Hubble Space Telescope had absolute highest priority for NASA, smaller satellites like ROSAT and COBE had to wait in line. At that time, the earliest possible ROSAT launch was indicated for 1994, a delay of seven years. Fortunately, it was possible to modify ROSAT for launch with a NASA Delta rocket. This only required removing the supporting structure for the shuttle bay and converting the solar cells to a folding mechanism. ROSAT launched on June 1, 1990.

COBE went through a similar procedure, but the satellite had to be substantially redesigned and its weight reduced so that it would fit on a Delta rocket. Fortunately, the instruments could be flown virtually as planned. On November 18, 1989, COBE finally launched. A few months later, Mather presented the first measurement results of the FIRAS instrument at the annual meeting of the American Astronomical Society in Washington, based on a measurement time of only nine minutes. He received a standing ovation upon showing a curve with a total of seventy-six measurement points and absolutely tiny error bars that described the entire spectrum of microwave background radiation to the left and right of the maximum. The measurement curve perfectly fit the model of a blackbody spectrum that both Gamow and Peebles had predicted. The temperature could be precisely determined to be 2.735 K. Three years later, Mather presented the data for the entire COBE mission. The error bars this time were only 0.03 percent— so small that they could not be shown in the final image because the model was drawn at a line thicker than the error range. In Figure 3.2, these error bars are therefore magnified by a factor of four hundred. The microwave background corresponds to a blackbody spectrum with the highest precision, and COBE determined its final temperature to be 2.726 ± 0.01 K. This therefore constitutes an essential pillar of the hot Big Bang model!

At the 2006 International Astronomical Union's general assembly in Prague, Czech Republic, where Pluto was demoted to a "dwarf planet" (see Chapter 6), the entire COBE team received the Gruber Cosmology Prize. A few months later, Mather and George Smoot were awarded the 2006 Nobel Prize in Physics.

THE "FACE OF GOD"

In addition to spectral information, the spatial distribution of radiation over the sky contains important information about the period of inflation shortly after the Big Bang. Earlier measurements have already shown that the background radiation varies by less than 1 percent in different directions. Using the DMR, COBE succeeded in measuring the distribution of radiation over the sky at three different wavelengths with very high precision. Plate 1 shows COBE's measurements of the intensity distribution across the sky at a frequency of fifty-three gigahertz in a false-color representation. The blue areas have a slightly higher temperature than the red areas. In the upper part of the figure, a global anisotropy shows a dipole variation across the sky. The blue regions on one side have a temperature approximately one-thousandth higher than average while the red side has a correspondingly lower temperature than average. The movement of the solar system at a speed of about 370 kilometers per second relative to the microwave background explains this anisotropy. Interestingly, the direction of motion derived from these measurements does not coincide with the rotation of the solar system around the Galactic Center but points almost in the opposite direction. This means that our galaxy and the whole Local Group of galaxies move against the rest frame of the microwave background at a speed of more than 600 kilometers per second. This movement is due to the superposition of different velocity components: the rotation speed of the sun around the Galactic Center is 220 kilometers per second, and the Milky Way moves at about 50 kilometers per second within the Local Group, which in turn travels at about 200 kilometers per second toward the Virgo Cluster, the largest galaxy cluster in the solar neighborhood. This, on the other hand, moves in the direction of the so-named Great Attractor (see Chapter 4). Thus, the microwave background represents a kind of absolute, nonmoving reference system.

Subtracting the dipole distribution due to the sun's motion from the background, we obtain the representation in the center of Plate 1.

Like the image of the Milky Way in Figure 1.1, this figure is in galactic coordinates with the Galactic Center in the middle. One can now clearly see the emission of the Milky Way, which originates from the synchrotron radiation of high-energy electrons in the galactic magnetic field. By subtracting a model of the Milky Way radiation from the data, we finally obtain the lower graph of Plate 1. The differences between the peaks and valleys now only correspond to about thirty microkelvin, or one-hundred-thousandth of the average temperature.

The microwave background, the echo of the Big Bang, is isotropically distributed across the sky with the utmost precision—as it should be if the hot radiation comes from the Big Bang. But can that be right? When we observe the background radiation in a certain direction, we see a wall of fire that moves away from us at umpteen times the speed of light. The light corresponds exactly to a blackbody spectrum that was originally emitted at a temperature of 3,000 K and now reaches us at the temperature of 2,735 K. If we view the sky in the opposite direction, then the firewall also moves at umpteen times the speed of light but in the opposite direction. The areas of the universe that move with superluminal speed away from each other have no possibility of ever communicating. If one side sends out a light beam, it will never reach the other because both move apart faster than light can travel. Under these circumstances, the universe is unable to reach an all-encompassing thermal equilibrium. It is therefore all the more surprising that the universe, particularly the microwave background radiation, is exactly equal in all directions within a fraction of one-hundred-thousandth. The theory of inflation solves this so-called horizon problem brilliantly because prior to the inflationary phase, the universe was much smaller, and all areas of the universe stayed in constant contact so that a thermal equilibrium could form. Inflation did not destroy this thermal equilibrium. On the contrary, inflation "froze" the thermal equilibrium once and for all so that our visible universe has remained almost exactly isotropic.

The theory of inflation, however, made an important prediction: the quantum fluctuations of the vacuum, which must have reigned shortly before the inflation, should have had a very characteristic, so-called scale-invariant distribution. In this distribution, all the different wavelengths contribute equally to the fluctuations so that if you display the power of the fluctuations per wavelength interval versus wavelength you obtain a horizontal line. US physicist Edward Harrison and Russian cosmologist Yakov B. Zeldovich independently postulated this power spectrum in the 1970s to explain the formation of galaxies. It is therefore called the Harrison-Zeldovich spectrum. Inflation theory can derive the shape of this spectrum from the original quantum fluctuations blown up by a huge factor during the inflation phase of the early universe. These fluctuations transformed into real density variations that must have affected all components of the primordial soup in the same way. As a result, certain areas in the universe contained a bit more matter and had higher temperatures than those places where matter was missing. Later, all other structures formed from these density fluctuations.

It represented a triumph not only for the Big Bang model but also for the theory of inflation when on April 23, 1992, Smoot, scientific director of the COBE DMR experiment working at the University of California at Berkeley, presented the first image of the COBE-measured fluctuations and the derived power spectrum, which fell completely in line with the Harrison-Zeldovich spectrum.[10] COBE had a relatively coarse angular resolution of about seven degrees. The full moon, in comparison, has a diameter of half a degree. The COBE map (Plate 1, bottom) could therefore divide the sky into only about eight hundred individual cells and was still so noisy that one could derive the significance of the measurements only from a statistical analysis. In the subsequent press conference, Smoot got carried away by his enthusiasm and stated that he could discover the "face of God" in the COBE map. This statement made front-page headlines in several important newspapers, and at that time most of Smoot's colleagues still considered it to be too brash and inappropriate.

THE SOUND OF THE FIREBALL

The story is, however, far from complete. The form of the power spectrum, in particular at small angular intervals, contains abundant information about early cosmological evolution. Although areas in the universe having a greater extent than their local horizon were not causally linked and therefore maintained the primordial fluctuations until the recombination era, this did not apply to smaller regions. In areas of the sky smaller than one or two degrees, the photon-baryon fluid and the dark matter could locally interact over time and thus alter the originally simple scale-invariant fluctuations. The simplest interaction took place in the form of sound waves. In 1955, famous Russian physicist, dissident, and Nobel Peace Prize winner Andrei Dmitrievich

Sakharov dealt with the density fluctuations in an expanding, hot universe. Together with his teacher and mentor Zeldovich, in 1969 my Russian colleague Rashid Sunyaev calculated that these fluctuations must have generated vibrations, or acoustic oscillations.[11] Jim Peebles at Princeton University soon arrived at very similar results.[12] An overdense region in the early universe did not collapse immediately under its own gravity until it entered its particle horizon, within which each point causally interacted with all other points. The disturbance then started to collapse until the effect of gravity resulted in increased pressure and thus a higher temperature. The corresponding enhanced radiation pressure, in turn, had a feedback effect on the matter and pushed out the excess material until the game started again, in an oscillating motion. Like drumbeats, the pressure and the temperature changes spread as sound waves in the hot fireball. One can therefore say with complete justification that the radiation-dominated universe had its own "sound." The speed of sound, however, was about 57 percent of the speed of light!

From this sound, we can learn a lot about the properties of the oscillating matter. We know, for example, that our sun and other stars, which are also fireballs, are subject to a large amount of acoustic vibrations. Analyzing these tells us about their inner structures and compositions, in a manner similar to earthquake waves. You can also learn something about the size of a system from its sound: depending on whether you blow over a small or a large bottle, for example, the pitch of the tone sounds deeper or less deep.

When matter and radiation separated from each other during recombination, the fluctuations present in the hot Big Bang plasma were released and virtually frozen into the radiation. We should therefore expect to observe the corresponding fluctuations in the image of the microwave background. For large scales, we have already seen the original fluctuations in the COBE data with a flat power spectrum. For scales smaller than about one to two degrees, the acoustic oscillations build up at certain frequencies so that the fluctuations' amplitude increases relative to the primordial, scale-invariant power spectrum. This effect produces a number of characteristic maxima, Doppler peaks, in the power spectrum. As Sunyaev and Zeldovich calculated, the first peak at the largest angular scales corresponds to the sound horizon on the surface on which the light was scattered for the last time. This represents the exact distance traveled by the sound since the fluctuations decoupled after the inflation.

A determination of the angular scale of the first acoustic peak in the sky (see Figure 3.3)[13] relates this physical distance to a cosmic distance and yields a geometric measurement of the curvature of space. For a flat universe, as we know from the geometry we learned in school, the first acoustic peak is expected at an angular scale of about one degree. From the position of the first Doppler peak in the fluctuation spectrum, we can thus directly calculate the total energy and mass density of the universe—that is, the sum of the matter density and the dark energy density ($\Omega_{tot}=\Omega_m+\Omega_\Lambda$). The amplitude of the first peak is directly related to the Hubble constant and the baryon density and therefore allows statements similar to, but completely independent of, those obtained from primordial nucleosynthesis. In addition to the fundamental frequency of the acoustic oscillation are the corresponding harmonics with the double, triple, etc., frequency. The recombination spread slowly like a clearing fog, strongly damping the higher-frequency oscillations. This effect is called "Silk damping" after English astrophysicist Joseph Silk, who published *The Infinite Universe*[14] a few years ago. The damping acts similarly to turning down all the treble in a hi-fi system and only reinforcing the bass. As you can see, the power spectrum (Figure 3.3) is comparable to the graphic equalizers in modern hi-fi systems, which display the amplitude of bass, midrange, and treble on separate LEDs.

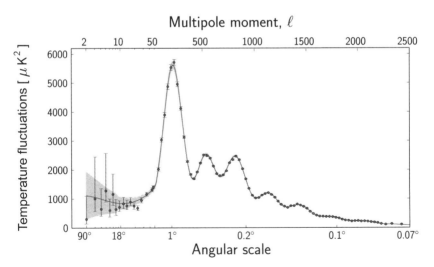

Figure 3.3. The cosmic microwave background (CMB) power spectrum determined from the Planck mission. (Courtesy of the European Space Agency and the Planck team.)

Immediately after the COBE team discovered the fluctuations in the cosmic background radiation in 1992, preparations began worldwide to hunt for the acoustic peaks at very small angular distances in the sky. For all scientists involved, it was clear this required a new long-term space mission using an all-sky survey with better angular resolution and sensitivity. As part of its Horizon 2000 program, the European Space Agency (ESA) solicited proposals for its third medium-sized space mission, M3, in November 1992 and subsequently selected a mission to study the microwave background at a high angular resolution. The satellite would carry a telescope with a diameter of 1.5 meters to measure the sky in nine channels with wavelengths from 0.3 to 12 millimeters at an angular resolution about fifty times better than COBE's.[15] After several iterations and difficult years in the ESA program, this satellite was renamed the Planck Surveyor in honor of the great physicist Max Planck. In 2008, it was launched along with ESA's Herschel Observatory cornerstone mission—sixteen years after the original proposal.

NASA has significantly shorter decision periods in its Explorer program, which in this case made a great impact in its favor. In 1994, a group under the leadership of Charles L. Bennett of the Goddard Space Flight Center proposed to build a satellite known as the Microwave Anisotropy Probe (MAP).[16] This group also included David Wilkinson, one of the original Princeton pioneers in the search for the "echo of the Big Bang." A few months before ESA announced its decision to build the Planck Surveyor, NASA decided to follow Bennett's competitive proposal and prepare MAP for a launch in 2001. The satellite, which carries passively cooled microwave radiometers with 1.5-meter reflectors, measures the microwave background at five different wavelengths ranging from 3.3 to 13.6 millimeters with an angular resolution about twenty times better than COBE's. The satellite was launched in June 2001 on a Delta rocket. After Wilkinson's sudden death in September 2002, NASA renamed this satellite the Wilkinson Microwave Anisotropy Probe (WMAP) and announced the first measurement results in February 2003.

Because of the relatively long development timescales for space experiments, the first breakthrough came from ground-based and balloon-borne microwave telescopes. As already described, measurements in the millimeter range in the vicinity of maximum background radiation are very difficult to obtain from the ground. But there are very dry areas in the world with minimal water vapor in the atmosphere, such as the Atacama Desert, a high plateau in the Chilean Andes, and the icy but very dry Antarctic. In addition, the technology of microwave receivers has become much more sensitive, especially through the use of new amplifiers that work with so-called High-Electron-Mobility Transistors (HEMT) and cool to within a few thousandths of a degree above absolute zero. A group of scientists that included Wilkinson achieved the first measurement of fluctuations to angular scales of about one angular degree in 1993. They worked with a Mobile Anisotropy Telescope (MAT) in the Canadian city of Saskatoon.[17] The same telescope was used in the Chilean Atacama desert in two campaigns in 1997 and 1998 on the 5,200-meter mountain Cerro Toco. In summer 1999, this group published the first evidence of the Doppler peak at about one degree in the fluctuation spectrum of the background radiation.[18]

Flights with high-altitude research balloons are a relatively inexpensive and fast development alternative to satellite missions but have the disadvantage of allowing only very short measurement durations. Because the jet stream carries the balloons at an altitude of forty kilometers away, flights over the entire American continent or Australia generally last only a few hours up to a maximum of two days. In Antarctica, however, long-term flights are possible in which the balloon, borne away by the wind, circles the South Pole along a particular geographic latitude and thus records measures for many days. In 1993, an international group under the direction of Paolo de Bernardis of the University La Sapienza in Rome prepared a balloon experiment called Balloon Observations of Millimetric Extragalactic Radiation and Geophysics (BOOMERanG). The payload consisted of a 1.3-meter reflector telescope and four receivers for microwave wavelengths in the range of 0.75 to 3.3 millimeters. Just a brief test flight over North America in 1997 indicated a peak in the angular power spectrum at about one degree and thus independently confirmed the Cerro Toco experiment. For the first long-duration flight in December 1998, BOOMERanG was launched from the McMurdo Station in Antarctica and drifted approximately along the 79th southern latitude. After a total of 259 hours of continuous measurement time, the balloon had circumnavigated the South Pole and floated back down to the ground only fifty kilometers away from its launch site. During the flight, the instrument mapped a thirty- by sixty-degree area in the southern sky. In April 2000, the BOOMERanG team published the data from this flight. The measurement results showed the first Doppler peak with unprecedented accuracy—the root note in the sound of the microwave background.[19] From the position of the peak, a value of $\Omega_{tot} = 1.01 \pm 0.06$ could be determined, consistent with a flat geometry for the universe, to within an accuracy of 6 percent.

The publication of the WMAP data in February 2003 can be considered a climax and a triumph for the inflationary Big Bang theory. Plate 2 (middle) shows the complete Wilkinson probe map after subtracting the impact of our motion relative to the background and the emission of the Milky Way, in direct comparison to the COBE map. At the first sight of this image, I was impressed not only by its excellent angular resolution but also by the fascinating fact that the WMAP has confirmed practically all structures discovered by COBE. This is the universe's baby picture 380,000 years after the Big Bang! Analysis of the WMAP's power spectrum shows that the small red, yellow, and blue spots represent not measurement noise but real signals from the sky. Almost a decade later, the first cosmic microwave map from the Planck Surveyor was published (see Plate 2, bottom), which provides the utmost angular resolution and wavelength coverage and is now the new gold standard for the microwave background. Figure 3.3 shows the cosmic microwave background angular power spectrum derived from Planck satellite data alone. Apart from the fundamental tone, one can see the higher harmonics of the Doppler peaks and the Silk damping that so beautifully confirm the "sounds of the Big Bang" predicted decades ago. The position of the first peak can be certified to approximate a flat universe with amazing accuracy: $\Omega_{tot} = 1.0023 \pm 0.0056$ (see Table 1 in the Appendix). Also, the scalar index n_s of the primordial fluctuation spectrum is very accurate at the scale invariant value—slightly less than the 1.0 predicted by Harrison and Zeldovich. These measurements strongly confirm the predictions of inflation theory. Cosmology has become a true precision science in recent decades.

In addition, the Planck Surveyor should be able to measure the polarization of the microwave background radiation with a precision never seen before. This in turn allows new diagnostics regarding gravitational waves produced directly at the time of the Big Bang. Actually, a press release about the Background Imaging of Cosmic Extragalactic Polarization 2nd Generation (BICEP2) team caused considerable excitement in early 2014. BICEP2 was a twenty-six-centimeter microwave telescope with a very large detector array, situated at Earth's South Pole. The team claimed to have found the B-mode polarization in the cosmic microwave background, which could be due to gravitational waves produced during the inflationary phase of the universe. However, after considerable debate in the community, the authors downgraded their significance in the official publication. In a joint analysis of the BICEP2 and the Planck satellite data the claim was formally withdrawn.

Chapter Four

THE COSMIC WEB OF GALAXIES

THE SEVEN SAMURAI AND THE GREAT WALL

In the 1970s, the renowned Jim Peebles, one of the great theoretical astrophysicists of our time whom we already got to know in the context of cosmic microwave background radiation, tried to understand the formation of the large-scale structures of the universe. He became preoccupied with how, in a relatively short period of time, the legions of galaxies and galaxy clusters with density concentrations of about one thousand to one (i.e., extremely clumpy) had been able to form out of the still uniformly hot fireball, with density fluctuations of about one to one hundred thousand (i.e., extremely smooth), at the time of the decoupling of radiation and matter. At that time, scientists knew how the galaxies were distributed from photographic sky surveys of the 1950s and 1960s, such as the *Shane-Wirtanen Catalog,* which showed more than one million counted galaxies using astrographic plates from the Lick Observatory (see, e.g., Figure 4.1). Additionally, Fritz Zwicky had obtained a catalog of thirty-one thousand galaxies from the photographic plates of the Palomar Sky Survey.

When looking at the distribution of the galaxies, you get the impression of a kind of foamy structure. Peebles and his colleagues mathematically analyzed this data and found clues suggesting a filament-like structure. But because no three-dimensional information was available, they could not be sure if it was due to projection effects or other optical illusions.

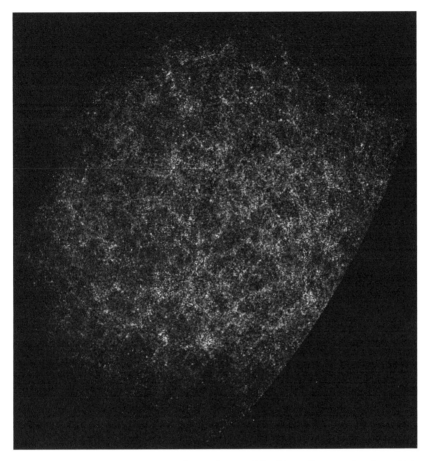

Figure 4.1. The approximately one million galaxies of the *Shane-Wirtanen Catalog*, which were counted on astrographic plates from the Lick Observatory. (Courtesy UC Regents/ Lick Observatory.)

Three-dimensional information about how the galaxies are distributed can only be obtained by spectroscopy. As Hubble had already shown, the recession velocity of galaxies increases with their distance. It is therefore possible to turn this relation around and estimate the distance of a galaxy from the redshift in its spectrum. After the pioneers Vesto Slipher, Milton Humason, and Gérard de Vaucouleurs had been observing galaxies for several decades on the biggest telescopes, about one thousand redshifts became known across the whole sky; far too few to say anything about the three-dimensional structure. In the mid-1970s, a technological revolution occurred in the field of astronomy when the first electronic detectors became commercially available in the form of big image-intensifier tubes. These image intensifiers had about twenty times the light intensity of the best photographic plates so that overnight, a small one-meter telescope became just as powerful as the biggest telescope of the time, the five-meter mirror on Mount Palomar. At this point, John Huchra becomes involved. As a postdoctoral researcher at the newly established Harvard-Smithsonian Center for Astrophysics (CfA) in Cambridge, Massachusetts, he met Margaret Geller and Marc Davis, who had both worked at Princeton University with Peebles. Along with students and engineers, the group built their own spectrograph and used the 1.5-meter telescope on Mount Hopkins to carry out the CfA Redshift Survey, the first substantial one of its kind. Davis, Huchra, Dave Latham, and my Hawai'i colleague John Tonry ran the initial CfA survey, which covered 2,400 objects, from 1979 to 1982. However, it still lacked the sensitivity to clearly depict the cosmic structure. Huchra, Geller, and Geller's doctoral candidate Valerie de Lapparent carried out a second survey (CfA2) between 1985 and 1995 that covered more than eighteen thousand galaxies.

Due to the very limited time available on the telescope, Geller convinced the group to perform the survey within a long strip in the sky in order to obtain the best data. Researchers have applied this trick to practically all modern redshift surveys. In his short autobiography, Huchra wrote:

> The problem was that there was a great debate about the methodology of the next survey. There were essentially three plans floating around. Marc Davis suggested a knitting needle approach, namely sampling one-in-five or one-in-ten of the fainter galaxies to increase the volume surveyed very rapidly, but not so densely. Simon White, another player in the game, wanted dense sampling but in a smaller, contiguous, square or rectangular area of the sky. Margaret was convinced that long and relatively thin strips across the sky were the way to go. You can think of the mapping problem this way. Suppose you want to study the topography of the surface of the Earth, and you have a steerable satellite but only a limited amount of film, say enough to take pictures of 1000 square miles. You could take random 1 square mile shots of the surface (the Marc Davis approach), you could carefully image a 33 × 33 mile square (the Simon White approach) or you could try to observe a strip, say 5000 miles by 1/5 of a mile (the Margaret Geller approach). The first approach would give you a pretty good idea of the fraction of the

Earth's surface covered by ocean, desert, mountains, etc., but you wouldn't know anything about the sizes of such things. This type of sparse sampling was actually used for one of the earlier IRAS galaxy redshift surveys, the QDOT (Queen Mary/Durham/Oxford & Toronto) survey, and produces a deep but very low resolution map, ok for continents, but watch out for mountain ranges! The second approach would give you very detailed information about a specific place, but since you're likely to see only ocean or desert or mountain, you'd have a very distorted view of the Earth. The third approach, however, is a winner, since not only are you likely to cross a little bit of everything, but you can also estimate the sizes of the oceans, continents and mountain ranges you cross. Not a map, but surely a mapmaker's first crude topographical survey![1]

When they plotted the data of the first CfA2 survey strip in 1985, Huchra and his colleagues were surprised at the distinct and diverse structures in the redshift map (Figure 4.2). The distribution of galaxies looked like a "hanged man." At first, Huchra was shocked and tried to find an error, but the data stood up to all examinations. In the middle of the map, you can see the Coma Cluster, which appears as an elongated, radial structure due to the galaxies' peculiar high velocities.

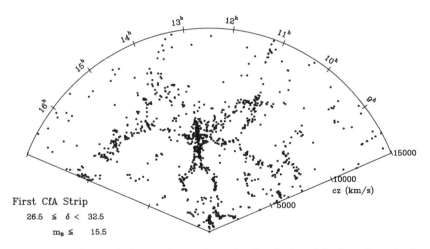

Figure 4.2. Wedge graph showing the two-dimensional projection of the three-dimensional galaxy distribution from the Harvard-Smithsonian Center for Astrophysics redshift survey. The distribution of galaxies is depicted as a wedge that is six degrees wide and about one hundred degrees long. The recession velocity of the galaxies is depicted along the radial dimension. (Courtesy Smithsonian Astrophysical Observatory.)

This phenomenon is also referred to as the "finger-of-God" effect. The long drawn-out line of galaxies that runs across the picture from east to west is called the Great Wall and remains one of the largest-known structures in the universe, measuring $600 \times 250 \times 30$ million light-years.[2] The galaxies are located along stretched-out filaments that surround vast cosmic voids. Four years previously, Bob Kirschner and his colleagues had discovered the first void of this kind, but for quite some time, it was not considered relevant. This changed abruptly with the existence of the CfA map. Geller was the first to coin the expression "soap-bubble universe," which still provides us with a vivid image of the Cosmic Web.

The discussion about the finger-of-God effect in Figure 4.2 demonstrates that redshifts do not enable us to distinguish between the galaxies' velocities caused by cosmic expansion and their proper motion relative to the surrounding galaxies. But if you manage to independently determine the distances of an adequate number of galaxies, the proportion of cosmic expansion can be eliminated, and the proper motion of the galaxies can be determined. At the same time as Huchra and his colleagues, another group of astronomers calling themselves the Seven Samurai were examining these so-called peculiar velocities of galaxies in nearby regions. David Burstein, Roger Davies, Alan Dressler, Sandra Faber, Donald Lynden-Bell, Roberto J. Terlevich, and Gary Wegner found, just like Huchra's group, that the galaxies are spread very irregularly through space, with superclusters of galaxies separated from one another by incredibly large voids of visible matter.

By analyzing the peculiar velocities of nearby galaxies and galaxy clusters, they discovered that these galaxies in the nearby universe must be attracted by a dark matter accumulation of about 5×10^{16} solar masses roughly 400 million light-years in size and located at a distance of about 250 million light-years in the direction of the southern constellation Centaurus. But unfortunately, the Great Attractor, as the Seven Samurai named it, conceals itself in the Zone of Avoidance in the sky, which blocks our direct view of roughly a fourth of the extragalactic cosmos due to the dense dust clouds and star fields of our Milky Way. It appears as though millions of galaxies are drawn to the Great Attractor at velocities from about six hundred (for the Local Group) to several thousand kilometers per second. These galaxies are located in a region that comprises the Milky Way, the Local Group of about thirty to forty nearby galaxies, the considerably larger Virgo Supercluster, and the nearby Hydra-Centaurus Supercluster. The

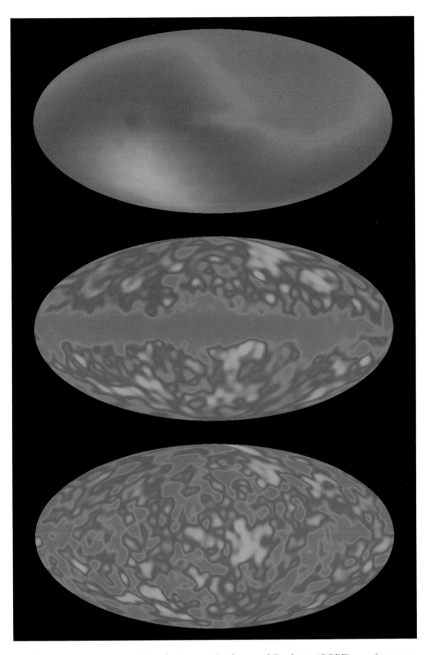

PLATE 1. Sky map measured by the Cosmic Background Explorer (COBE) at a frequency of fifty-three gigahertz in a false color representation. *Red* signifies a slightly higher temperature; *blue* a correspondingly lower temperature. The dynamic between the maximum and the minimum is about 1/1,000 in the top display. The dipole distribution of the temperature in the sky comes from the motion of the sun relative to the background. In the middle illustration, this dipole effect has been subtracted. Thus, the emission of the Milky Way is clearly visible. In the lower diagram, a model of the Milky Way emission has been subtracted so that the cosmic fluctuations are visible. The dynamics of the temperature scale here is about 1/100,000. (Courtesy of NASA and the COBE team.)

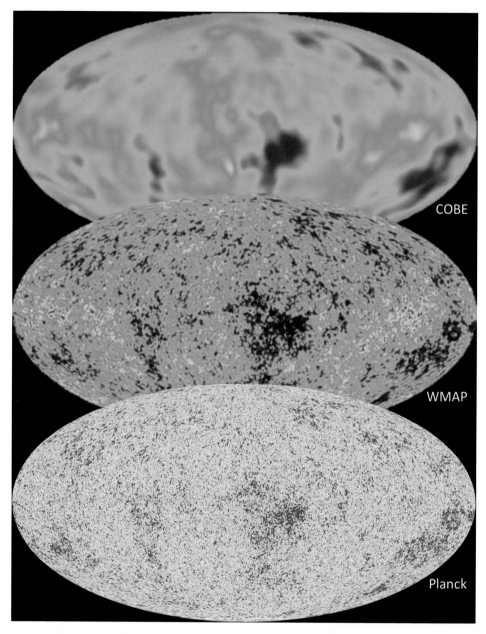

COBE

WMAP

Planck

PLATE 2. *Top:* The Cosmic Background Explorer (COBE) map of the cosmic microwave background. (Courtesy of NASA and the COBE team.) *Middle:* The map measured by the Wilkinson Microwave Anisotropy Probe (WMAP). (Courtesy of NASA and the WMAP team.) The main structures of the COBE map were confirmed by WMAP—however, with much better angular resolution and sensitivity. *Bottom:* The map from the first year of Planck Surveyor data with yet higher resolution, sensitivity, and wavelength coverage. (Courtesy of the European Space Agency and the Planck team.)

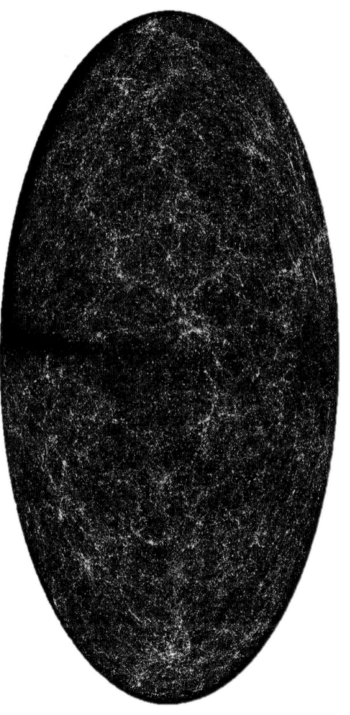

PLATE 3. Galaxies in the infrared sky. Closer and more distant structures are color-coded between blue and red to create a three-dimensional effect. (Courtesy of the California Institute of Technology and the University of Massachusetts.)

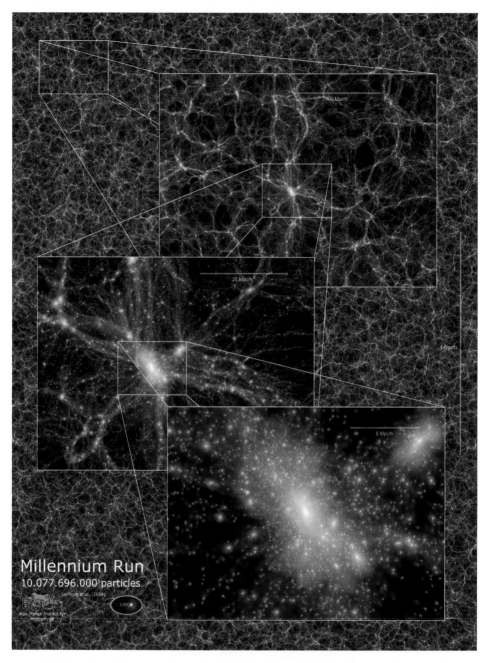

PLATE 4. The Millennium Run of the VIRGO Consortium with over ten billion matter particles. The picture shows a 15 Mpc/h section of the simulation at a redshift of $z = 0$. Each of the overlaid images zooms in with a factor of four; the areas marked by the white rectangles are enlarged and the scales are given. The number of particles per galaxy is similar to that in Figure 4.4. (Courtesy of Volker Springel, Heidelberg University.)

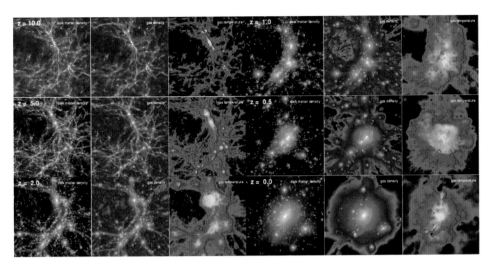

PLATE 5. The high-resolution simulation of the formation and evolution of a massive galaxy cluster. Within the cluster, more than twenty million mass particles are simulated. For every step in time indicated by the redshift (*top left*), the distribution of dark matter (*left*), the distribution of normal baryonic gas (*middle*), and the temperature of the gas (*right*) is shown. The temperature scale is color-coded and includes values of up to ten million K. You can see how the original filamentary structure of the Cosmic Web gradually turns into a compact spherical structure filled with hot gas. (Courtesy of Volker Springel, Heidelberg University.)

PLATE 6. The Hubble Ultra-Deep Field (HUDF), taken with the Hubble Space Telescope. This is the most sensitive picture of the sky ever taken with an optical telescope. It is about three arcminutes in size (one-tenth the diameter of the full moon) and contains nearly ten thousand galaxies. The picture was put together from eight hundred single images obtained by the Hubble telescope while orbiting Earth over four hundred times over a span of 11.3 days. (Courtesy of NASA; the European Space Agency; S. Beckwith and the HUDF Team/Space Telescope Science Institute [STScI]; and B. Mobasher of the STScI.)

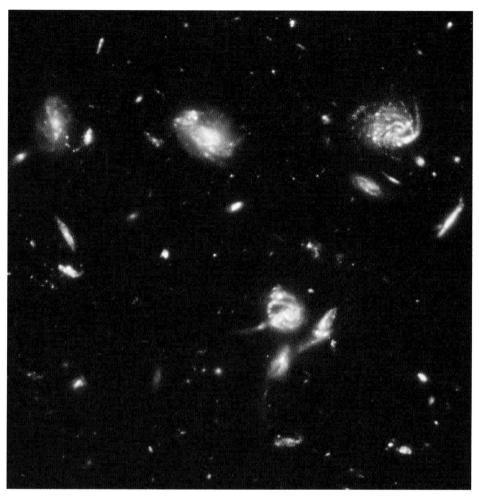

PLATE 7. Section of the Hubble Ultra-Deep Field. (Courtesy of NASA; the European Space Agency; S. Beckwith and the HUDF Team/Space Telescope Science Institute [STScI]; and B. Mobasher of the STScI.)

PLATE 8A; 8B. The Sombrero Galaxy, which owes its name to a prominent dust disc in the outer regions. This "in-between" type of elliptical and spiral galaxy lies at a distance of about fifty million light-years and is roughly the size of our Milky Way. *Top:* Image taken with the Hubble Space Telescope, in which the colors are more or less equivalent to visual perception. (Courtesy of NASA and the Hubble Heritage Team, Space Telescope Science Institute/Association for Universities for Research in Astronomy (STScI/AURA). The galaxy mostly consists of old, red stars; the dust disk appears dark in front of the bright background of stars. *Bottom:* A combination of the optical image (*blue*) and an infrared image of the Spitzer Space Telescope. Here the dust disk has a red glow because the cold dust mainly emits at long wavelengths. (Courtesy of NASA/Jet Propulsion Laboratory–California Institute of Technology and the Hubble Heritage Team, STScI/AURA.)

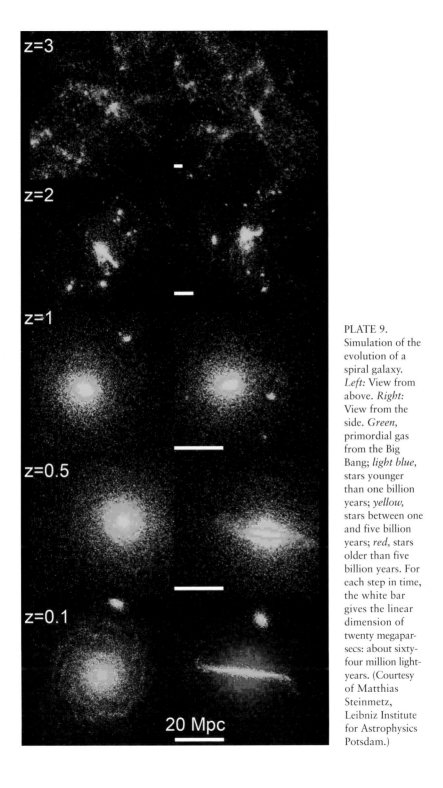

PLATE 9. Simulation of the evolution of a spiral galaxy. *Left:* View from above. *Right:* View from the side. *Green,* primordial gas from the Big Bang; *light blue,* stars younger than one billion years; *yellow,* stars between one and five billion years; *red,* stars older than five billion years. For each step in time, the white bar gives the linear dimension of twenty megaparsecs: about sixty-four million light-years. (Courtesy of Matthias Steinmetz, Leibniz Institute for Astrophysics Potsdam.)

PLATE 10. *Above:* This wide-angle image was taken with the 0.9-meter telescope on Kitt Peak. (T. A. Rector, University of Alaska Anchorage, NRAO/AUI/NSF and NOAO/AURA/NSF, and B. A. Wolpa, NOAO/AURA/NSF.) It shows the famous Eagle Nebula M16, a star-forming region in the Serpens constellation. Highlighted in green are the interesting "elephant trunks," which were captured in 1995 by the Wide Field and Planetary Camera (WFPC2) and in 2004 by the Advanced Camera for Surveys (ACS) of the Hubble Space Telescope (see also Plate 11).

PLATE 11. *Opposite page:* A section of the Eagle Nebula M16 taken with the Hubble Advanced Camera for Surveys (ACS) camera. (Courtesy of NASA; the European Space Agency; and the Hubble Heritage Team, Space Telescope Science Institute/Association for Universities for Research in Astronomy.)

PLATE 12. *Top:* Images of the Bok globule Barnard 68 in visible and in infrared light. *Left:* Optical image taken with the European Southern Observatory's (ESO) 8.2-meter Very Large Telescope (VLT) with the FORS1 instrument. *Right:* False color image of the optical image, *blue* and *green,* and an image in the infrared range, *red,* with the SOFI instrument of the ESO's 3.6-meter New Technology Telescope (NTT) in La Silla. (Courtesy of ESO.)

PLATE 13. *Bottom:* The Herbig-Haro object HH 46/47 observed with the Atacama Large Millimeter/submillimeter Array (ALMA) and in visible light from the European Southern Observatory's (ESO) New Technology Telescope. *Lower right:* The ALMA observations, in *orange* and *green,* of the newborn star reveal a large energetic jet moving away from us, which in the visible is hidden by dust and gas. *Left:* In *pink* and *purple,* the visible part of the jet is seen, streaming partly toward us. (Courtesy of ALMA, (ESO/NAOJ/NRAO)/ESO/H. Arce. Acknowledgments: Bo Reipurth (IfA, University of Hawai'i.)

Great Attractor is also pulling in galaxies from the far side of the universe that move a little slower compared to the cosmic expansion of the rest of the universe. A catalog of galaxy clusters has also been identified in the Zone of Avoidance using the Roentgen Satellite (ROSAT) All-Sky Survey.[3] Unlike visible light, X-rays can penetrate the clouds of gas and dust in our Milky Way, making it possible, for the first time, to measure the distribution of matter heading toward the Great Attractor.

University of Hawai'i at Mānoa astronomer R. Brent Tully and his international team of astronomers recently developed a new method to identify and localize superclusters by mapping their effect on the motion of galaxies. By very accurately mapping the location and the velocities of thousands of galaxies throughout our local universe and thus the gravity of the large-scale structure responsible, the team defined the regions of space dominated by different superclusters. According to this study, the Milky Way resides in the outskirts of one such supercluster named *Laniakea,* meaning "immense heaven" in the Hawai'ian language. The name honors Polynesian navigators who used their knowledge of the heavens to voyage across the immensity of the Pacific Ocean.

The Laniakea Supercluster is 500 million light-years across and contains about 10^{17} (one hundred quadrillion) solar masses, or suns, in some one hundred thousand galaxies.[4] After thirty years, this work finally clarifies the role of the Great Attractor. Within the volume of the Laniakea Supercluster, the galaxy "motions are directed towards the center, as water streams follow descending paths toward a valley. The Great Attractor region is a large flat bottom gravitational valley with a sphere of attraction that extends across the Laniakea Supercluster."[5]

After Huchra completed the CfA Redshift Survey, he initiated another big-sky survey project, the Two Micron All-Sky Survey (2MASS).[6] It took advantage of the penetration power of infrared radiation, as well as the fact that most normal star populations have an emission maximum in the near-infrared range. The University of Massachusetts, the CfA, and the Infrared Processing and Analysis Center (IPAC) of the California Institute of Technology carried out the project. Between 1997 and 2001, it used 1.3-meter telescopes on Mount Hopkins in Arizona and at the Cerro Tololo Inter-American Observatory in Chile to map the entire sky in the infrared range. The final 2MASS catalog contains more than 470 million objects, which include over 1.6 million nearby galaxies. In Plate 3, these galaxies are color-coded according to their luminosity, creating a three-dimensional effect that nicely illustrates the filament-like structure of the Cosmic Web. Due to the higher penetration power of infrared radiation, the galactic plane in the center of the image can be vaguely recognized.

In the meantime, the technology used for big-sky surveys has advanced dramatically. The Two-degree Field Galaxy Redshift Survey (2dFGRS), which obtained the spectra of nearly 250,000 objects, was performed in the southern sky using a large wide-angle spectrograph on the four-meter Anglo-Australian Telescope (AAT). One of the most ambitious astronomical projects so far has been the Sloan Digital Sky Survey (SDSS), which has obtained detailed optical images of more than a quarter of the entire sky, as well as a three-dimensional map of about one million galaxies and quasars. Since the beginning of this survey, the data from this worldwide community of astronomers have been made available annually. The SDSS uses its own 2.5-meter telescope, which is located on Apache Point in New Mexico and is equipped with two special instruments: a 120-megapixel camera that covers an area of the sky three times the diameter of the full moon and a double spectrograph with optical fibers that enable it to observe more than six hundred galaxies with every single observation. The SDSS's first survey, SDSS-I, was completed in June 2005. It captured about 200 million objects in five colors and the spectra of approximately 950,000 objects covering more than eight thousand square degrees. In Figure 4.5 (top), you can see that the SDSS revealed a wedge section of the large-scale structure that now, compared to the CfA survey, appears much more vivid and distinct. Once again, you can see the Great Wall of the filament nearest to us, packed with galaxies. The SDSS gathers a stream of data so immense (many terabytes) that even the catalogs are not handed out to all astronomers anymore. Instead, many scientists inquire directly via the survey's Internet-based computer server. The fact that the data are growing increasingly larger and more diverse has led the worldwide community to develop a "Virtual Observatory." Ideally, users no longer need bother with where the data they are interested in are stored—they simply let the Virtual Observatory sift through all the archives for information.

Cosmology and Killer Asteroids

The race of superlative astronomical surveys has yet to come to an end. The next huge astronomical survey project is the Panoramic Survey Telescope and Rapid Response System[7] (Pan-STARRS) that my Hawai'i colleague Nick Kaiser conceived. Pan-STARRS is an array of wide-angle astronomical telescopes with gigapixel CCD (Charge Coupled Device) cameras and a powerful computing facility that produces a kind of color movie of the whole northern sky. The first prototype telescope of the Pan-STARRS project, the PS1, was constructed on top of the volcano Haleakala on the Hawai'ian island Maui. It is a 1.8-meter telescope with a totally new wide-angle design and a 1.4-gigapixel camera, the largest ever built. This camera, developed by John Tonry, made the list of the twenty engineering wonders of the world. If you have a camera in your mobile phone, it may have about ten megapixels. Imagine 140 smartphones taking simultaneous images every minute through a two-meter lens and then you can imagine the tremendous data stream that this system produces—about a terabyte per night. It is as if the astronomers were drinking from a fire hose. The PS1 has completed its first mission, funded by the international PS1 Science Consortium, which continuously surveyed three-quarters of the whole sky and a number of deep fields over four years. The accumulated images are significantly deeper and cover a substantially larger area than the SDSS data. The final data is to be made public starting in summer 2015 under the leadership of my colleagues Ken Chambers and Gene Magnier. Currently, the second Pan-STARRS telescope, the PS2, is being finalized for operation on Haleakala (see Figure 4.3).

Figure 4.3. The night sky above Haleakala on Maui with the Pan-STARRS Observatory. The second telescope, PS2, is in the foreground, partially occulting the PS1 dome. (Courtesy of the University of Hawai'i.)

As director of the Institute for Astronomy (IfA) at the University of Hawai'i, I am responsible for the care of the Pan-STARRS project. One of its interesting aspects is its funding. The costs for its system development, the prototype telescope PS1, and a large fraction of the PS2 were borne by the US federal government and the funding was managed by the US Air Force. Why is the US military so interested in this "celestial census"? They worry about "killer asteroids," those celestial objects as big as mountains that, just like in the Hollywood movies *Armageddon* with Bruce Willis or *Deep Impact* with Robert Duvall, may potentially cause a mass extinction on Earth. One such impact at the end of the Cretaceous period is often held responsible for the extinction of the dinosaurs. In 1998, the US Congress mandated that NASA find 90 percent of all near-Earth objects larger than one kilometer, which could cause a mass extinction. According to a NASA press release, it reached this goal in September 2011 after discovering more than nine hundred objects—none of them on a collision course. However, in 2005 the congressional mandate was extended to finding 90 percent of all dangerous objects 140 meters or greater by the year 2020, and we are nowhere near this goal yet.

This is where Pan-STARRS and the air force enter the picture. As the mission of the air force is national defense and planetary protection, it is also concerned about possible "enemies" from outer space. Because the air force already operates a number of telescopes on Haleakala, it has been the prime partner for the IfA in administering the Pan-STARRS project. Due to the financial difficulties in Washington over the last few years, the completion of the PS2 project relied on funds from a variety of sources, including a major gift from an anonymous private donor, for which I am most thankful, and last but not least, NASA. For the next few years, NASA will also fund the operation of both Pan-STARRS telescopes.

I would like to take this opportunity to allow you and myself a brief digression concerning the detection of and the defense against killer asteroids. Again and again, we are startled by news about newly discovered potential killer asteroids that might hit Earth in umpteen years. After a couple of days, this is usually followed by the all-clear that the respective villain will, after all, scrape past Earth at a safe distance. This causes quite a number of people to get annoyed: Why do these astronomers make such a fuss instead of calculating properly in the first place? The problem lies in the inaccuracy of the measurements. The predicted impact parameter (in other words, the smallest distance when flying past Earth) still has very large errors shortly after detecting the asteroid, and Earth quite often lies within this radius. The longer the measurements continue, the smaller the error interval becomes until Earth disappears out of the potential impact area altogether. This is also the case with the 320-meter asteroid 99942 Apophis, which will fly past Earth at a distance of only thirty thousand kilometers on Friday, April 13, 2029. Unfortunately, as it passes it might take with it one of the geostationary satellites that orbit Earth at roughly this distance.

The statistics on asteroids that we know about allow us to estimate how often a meteorite of a certain size hits Earth and gauge its destructive force. Objects with a diameter of about three meters reach Earth roughly once a year but have little effect because they usually burn up in the atmosphere. Many fall into the ocean unnoticed or hit uninhabited areas. But once in a while, a meteorite crashes in a densely populated area, alarming the inhabitants and causing the press to celebrate. On the night of April 6, 2002, the "meteor of Bavaria" shot across the Alps—and the fireball burned up dramatically in Earth's atmosphere. The scientists of the Institute of Space Sensor Technology and Planetary Exploration of the German Aerospace Center (DLR) in Berlin-Adlershof look after the European Fireball Network.[8] This network consists of twenty-five video cameras in Germany, the Czech Republic, Slovakia, Belgium, Switzerland, and Austria that constantly keep an eye on the entire sky by means of a small semicircular mirror. Due to the detailed information from these seven cameras, astronomers at the Ondrejov Observatory in Prague, Czech Republic, were able to precisely calculate the bolide's trajectory and found that they were dealing with a chunk weighing between five hundred and six hundred kilograms that entered Earth's atmosphere at a speed of thirty kilometers per second and burned up at an altitude of between eighty-six and sixteen kilometers. Its remnants—all in all about twenty kilograms of rock—must have landed somewhere between Garmisch-Partenkirchen and Schwangau in Germany. On July 14, 2002, after several systematic search operations in difficult mountainous terrain, the DLR finally found the first meteorite fragment near the famous fairy-tale castle of Ludwig II of Bavaria—and therefore named it Neuschwanstein, after the castle. The chunk, weighing 1.75 kilograms, is magnetic and covered with a matte black fusion crust containing rusty spots, although it seems these only formed as a result of the deep snow that the fragment smashed into. According to the DLR press release:

The valuable extraterrestrial find is now being carefully analyzed, chemically and petrologically, so that it can be assigned to the correct meteorite class. What is particularly interesting here is a comparison with the "Pribram" meteorite, which 43 years ago on 7 April 1959 was also photographed by the Fireball cameras and which was later found in Czechoslovakia. The model calculations showed that the orbit of "Neuschwanstein" is almost identical to the one of "Pribram," which is unusual with meteorites. This discovery could indicate that a whole stream of meteoritic bodies exist, which were possibly created by a small asteroid breaking up. The laboratory work will hopefully reveal if the two meteorites indeed originate from the same parent body and how far in the past the breaking up of the parent body occurred. According to statistical considerations there must be about one billion similar meteorites within the stream, which, taken as a whole, amount to an asteroid with a diameter of approximately 600 meters.[9]

According to this analysis, it could therefore indeed be possible that another impact originating from the same stream will occur—perhaps an even slightly bigger one! Many years later, in 2013, graduate student Marco Miceli at the IfA observed in his dissertation that some of the famous meteor showers regularly appearing in the night sky actually share the same orbit as the near-Earth asteroids and may be connected to the breakup of larger asteroids.

On June 30, 1908, an explosion in Siberia, not far from Tunguska, knocked down about sixty million trees over an area covering roughly two thousand square kilometers. Even at a distance of over five hundred kilometers, passengers of the Trans-Siberian Railway perceived a bright, fiery glow; a strong tremor; and a shock wave, as well as the sound of thunder. The nights that followed were so bright in parts of Europe that it was possible to read at night without any other light sources.[10] The explosion must have been roughly as strong as the most powerful hydrogen bomb ever detonated, the Tsar Bomba. This incident is ascribed to an asteroid with a diameter of about fifty to seventy-five meters. Impacts of this size are expected to occur approximately every two thousand to three thousand years, according to the Lincoln Near-Earth Asteroid Research (LINEAR) project.[11] Asteroids with a diameter of more than one kilometer hit Earth approximately every six hundred thousand years, according to LINEAR's findings, and have already caused great damage. Gigantic chunks with diameters of up to ten kilometers—for example, the Chicxulub asteroid, which is considered responsible for the extinction of the dinosaurs sixty-five million years ago—are expected to hit roughly every one hundred million years. In fact, a number of the dramatic mass extinctions in Earth's history have been linked to such impacts.

The most recent excitement resulted from the Chelyabinsk meteor, a fifteen- to twenty-meter diameter near-Earth asteroid that entered the atmosphere over Russia on February 15, 2013. It exploded at a height of about thirty kilometers with an explosion energy of five hundred kilotons of TNT or its equivalent: about twenty to thirty times the energy of the atom bomb dropped on Hiroshima. As such, it was the largest object to enter the atmosphere after the Tunguska event. The shock wave shattered many glass windows in the town of Chelyabinsk and injured more than one thousand people—fortunately, with no fatalities.

So what do we do if, someday, a killer asteroid takes a direct path toward Earth? Just as the previously mentioned notorious Hollywood movies epically demonstrate, the heroes would face great challenges: fly there and detonate a nuclear bomb? That would be the worst idea. Depending on the composition of the asteroid, a bomb would either fizzle out in the porous rock, or it would break the asteroid up into big fragments that nevertheless would still calmly head toward Earth. In actuality, no reliable method yet exists for diverting a killer asteroid from its course. Using an ion thruster to direct the asteroid from its path is a possibility, but it is risky because if any uncertainty exists about the exact location of the impact, the asteroid might be shifted in the wrong direction. Besides, most asteroids rotate, and if the rocket is not placed exactly in the center of gravity, it is more likely to change the direction of rotation rather than the course. Other methods for diverting meteorites were suggested a couple of years ago, including sending a powerful beam of concentrated sunlight to heat up and evaporate a small patch on an asteroid, changing its momentum and thereby altering its trajectory. Another idea is to throw the asteroid's heat absorption and emission off balance by painting one side white so that it gradually diverts from its course. Recently, the suggestion was made to tow it slowly from its path using the gravitational force of a really heavy "tractor spacecraft." But all of these techniques would have to be initiated very quickly to avoid disaster. And at the time in question, the trajectory would still be too uncertain to precisely predict. You can imagine that none of these methods are particularly suited to the drama of a Hollywood movie, in which success or failure rarely occurs until the final seconds.

Finally, asteroids have also captured the imagination and the interest of big business because some may contain precious rare elements that are believed to vanish on Earth in the near future. In 2014, NASA embarked on the idea to robotically capture a suitable asteroid and bring it down to an orbit around the moon by 2020, where astronauts could directly explore it. This bold vision recently replaced the idea to fly astronauts to Mars in the next few decades.

To Everyone Who Has Will Be Given

Apart from saving the Earth, the data from Pan-STARRS and other large sky surveys are a treasure trove for astrophysics and cosmology. This brings us back to the cosmos: Now, how are the dramatic cosmic structures formed—the filaments and the big voids, the galaxy clusters, and the superclusters? Going by our present understanding, the minimal fluctuations in the microwave background radiation represent the tiny differences in the density of the universe at the time of matter-radiation decoupling. These differences result from the quantum fluctuations during inflation and are therefore the first visible large-scale structures in the cosmos. In the end, it is the tiny differences in the density of matter that at first increase linearly under the force of gravity and later form the nonlinear large-scale structures. Because the gravitational force is attractive throughout, a small surplus of matter in one spot is sufficient to attract more matter from its surroundings and amplify the contrast between density and the force of attraction.

The Bible says, "For to everyone who has will be given, and he will have abundance, but from him who has not, even that which he has will be taken away."[12] This also applies to the gravitational force, which acts in a similar capitalist way. Tiny little matter surpluses gradually attract all of the matter to their vicinity. For the normal baryonic matter alone, this would require substantially more time than is available in the young universe after decoupling from the baryon-photon fluid 380,000 years after the Big Bang. Dark matter, on the other hand, along with the neutrinos, already decoupled much sooner than the baryons and can therefore collapse quickly enough. That way, by the time of decoupling, it has already formed the first filaments. Because the mass density of the dark matter by far outweighs that of the baryons, the latter simply get swept along in the collapse. The sum of dark matter and normal baryonic matter is responsible for the gravitational collapse and the deceleration of cosmic expansion, whereas dark energy reaccelerates cosmic expansion further.

Although it is not possible to directly observe the further destiny of the early universe, it can be simulated on modern supercomputers. Simulations are important to bring the observations regarding the radiation from the almost homogeneous fireball into accordance with the complex structures in today's universe. Figure 4.4 shows a simulation that looks solely at the development of dark matter.[13] It examines a cube-shaped section of the universe containing the right matter density and energy density in which the matter density is imprinted with the corresponding spectrum of the primordial fluctuations. Each of the five top pictures of Figure 4.4 shows this cube with an edge length of 140 million light-years (today), in which two million dark matter particles are left to their gravitational forces.[14] A typical galaxy consists of about seven hundred particles; for the sake of clarity, only 10 percent are depicted here. The numbers in the top-left corner of each picture stand for the corresponding redshift and therefore, vaguely, for the age of the universe at the respective point in time: $z = 28.62$ corresponds to approximately 100 million years, $z = 8$ corresponds to approximately 900 million years, and $z = 0$ corresponds to today's universe about 13.8 billion years after the Big Bang. The bottom row shows a considerably smaller section of the simulation for the same points in time but with all particles. In the first picture of this row on the bottom left, you can see quite well how the particles were initially

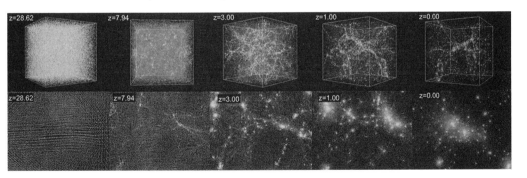

Figure 4.4. Simulation of the large-scale structure formation caused by the self-gravitation of dark matter from its homogeneous early phase. One galaxy corresponds to one of the compact clumps at the bottom right of the picture. The number at the top left of each picture indicates the redshift and therefore the momentary age of the universe. *Top row:* A cube with an edge length of 140 million light-years (today). *Bottom row:* A considerably smaller section of the same simulation. (Courtesy of A. Kravtsov, A. Klypin, and S. Gottlöber, Leibniz Institute for Astrophysics Potsdam.)

distributed almost uniformly in the box but with slight fluctuations. As if by an invisible hand, within a relatively short period the first condensation nuclei of bigger structures emerge from this matter like the whitecaps on breaking waves. After that, filaments form, from which the galaxies hang like a string of pearls, as well as big voids with hardly any galaxies at all. At the vortex of several filaments, dense areas with thousands of galaxies spring into existence—the galaxy clusters and the superclusters. In these dense areas, several galaxies very frequently merge to produce ever-larger structures. These distinctive structures are only obtained with the right amount of cold dark matter. If hot dark matter is used instead, all of the cosmic structures will be smeared out.

During the last decades, the processing power of supercomputers has increased dramatically throughout the world. Nevertheless, enormous simulations of the universe still present a great challenge that only the cooperation of several worldwide leading groups can master. The Virgo Consortium for Cosmological Supercomputer Simulations is an international group of scientists who use supercomputers to simulate the formation of galaxies, galaxy clusters, the large-scale structures, and the development of the intergalactic medium. Most members of the consortium are British, but important nodes also reside in the United States and Canada, as well as at the Max Planck Institute for Astrophysics in Garching, Germany, where my colleagues Simon White and Volker Springel[15] are in charge. The Edinburgh Parallel Computing Centre and the Computing Center of the Max Planck Society in Garching provide the primary computer resources.

Early in the twenty-first century, the Virgo Consortium published the biggest cosmological simulation ever conducted. Known as the Millennium Run, it used more than ten billion particles to observe the evolution of the matter distribution in a cube with an edge length of more than two billion light-years. The simulation kept the Max Planck Society's supercomputer busy for over a month (see Plate 4).

By applying cleverly devised methods of analysis to the roughly twenty-five terabytes of simulated data, the Virgo Consortium was able to portray the evolution of more than twenty million galaxies inhabiting an enormous volume in the simulated universe. In addition, they localized massive black holes, which sometimes make quasars light up in the center of galaxies (see Chapter 8). When directly compared with sky surveys, it is thus possible to study the physical

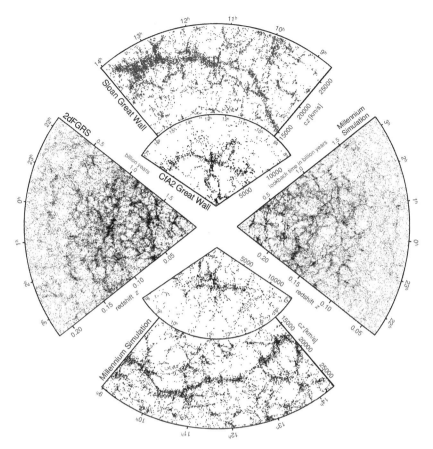

Figure 4.5. Comparison of sections from the Millennium simulation with the actually observed large-scale structures from the Harvard-Smithsonian Center for Astrophysics, the Sloan Digital Sky, and the 2dF Galaxy Redshift surveys. In the meantime, supercomputer simulations have become so advanced that it is almost impossible to distinguish between the virtual and the real sky. (Courtesy of Volker Springel, Heidelberg University.)

processes dominating the formation of galaxies and black holes in the real universe. How well the simulated sky matches the observations can be seen in Figure 4.5, which compares sections of the Millennium Run to similarly sized CfA redshift slice, SDSS, and 2dFGRS surveys. If the individual wedge graphs weren't marked, every astronomer would have difficulty discerning the virtual world from the real world. The simulated galaxies have been saved in a "theoretical virtual observatory" similar to the previously mentioned Virtual Observatory.

THREE-DIMENSIONAL TRAFFIC JAM

Today, supercomputer simulations have advanced so far that in addition to dark matter, they also take into account the gas physics of normal matter. The high-resolution simulation in Plate 4 shows the formation and development of a big cluster of thousands of galaxies emerging from the initial conditions of the cosmos.[16] This galaxy cluster contains more than twenty million mass particles and demonstrates the evolution of dark matter and hot intergalactic gas. The reason why I like this simulation so much is because, for the first time, it tells the story of the normal matter of which we all consist, of which we eat, and which we love. At the beginning ($z = 10$), the dark matter and the gas are located in the same three-dimensional filament-like structures that we already know from earlier cosmological simulations. The dark matter is in control; it is practically sitting in the driver's seat, whereas the normal matter is taken along for the ride.

By examining Plate 5 more closely, or better still, Volker Springel's video on the Internet, one gets the impression of a kind of three-dimensional road map. The lumps (halos) of dark matter drive like cars on filament streets toward the city; however, the streets themselves are likewise pulled into the city. The more tightly the whole bunch clumps together, the faster the dark matter particles move. The normal baryonic matter is, as mentioned before, at first swept along and is therefore distributed similarly to the dark matter. But on closer examination, one finds that the structures in the gas are smoothed out a little. In contrast to dark matter, which follows only gravity, the gas particles also obey electromagnetic interaction and can therefore radiate and cool.

When looking at the third column in the figure, we can observe that the matter in the dense areas is growing excessively hot. It begins releasing X-rays. While the matter is increasingly collapsing inward, the three-dimensional traffic jam causes the heat to spread outward through shock waves, and the gas heats up to such an extent that the galaxy cluster turns into an X-ray source of several million K. In a car traffic jam, vehicle brake lights and other visual effects send signals to drivers that very rapidly spread down the freeway, creating similar shock waves. In especially dense areas, the baryon clouds can efficiently cool down by emitting radiation. They then collapse under their own gravity and decouple from the dark matter. The cool, dense baryonic matter begins to form the first stars, black holes, and protogalaxies. The first light in the universe possibly arose at a redshift of about ten to twenty, corresponding to the very beginning of the simulation. We will hear more about this in later chapters. In today's universe, already more than half of the baryons are said to be hotter than 100,000 K; they can only be observed using ultraviolet or X-ray telescopes.

THE WHITECAPS

Galaxies—as far as the eye can see! The eye belongs to the famous Hubble Space Telescope, which, on March 1, 2002, during a visit from the crew of the space shuttle *Columbia,* was presented with a new imaging instrument—the Advanced Camera for Surveys (ACS). Due to its large field of view, its sharp pictures, and its improved sensitivity, this camera could perform deep surveys about ten times faster than its predecessor, the Wide Field and Planetary Camera (WFPC2).[17] In the summer of 2003, Steve Beckwith, who at the time was director of the Space Telescope Science Institute, asked if I'd be interested in becoming part of a scientific committee to help choose a location for the Hubble Space Telescope's most sensitive observations ever to be carried out in visible light. I happily accepted, especially because I had already worked on quite a few deep surveys in other wavelength ranges, mainly X-rays (see Chapter 8) and had realized the importance of coordinating the various teams' efforts. In the case of the Hubble Deep Field observation of the mid-1990s, this coordination had unfortunately not worked out too well. Today in the region of the Big Dipper, several deep fields of different wavelength ranges exist in immediate proximity to each other. As a result, they take observation time away from one another to some extent. The location for these newest observations, the Hubble Ultra-Deep Field (HUDF), was supposed to be ideal from the telescope's point of view while also optimally satisfying the basic conditions of other wavelength ranges—for example, the infrared astronomy, the radio astronomy, and the X-ray astronomy communities. For this reason, specialists on deep observations of all wavelengths, as well as representatives from the most important ground-based telescopes, were invited to join the ACS UDF Scientific Advisory Committee.[18] I was very pleased when, after several weeks of convincing, all members of the committee accepted my suggestion to place the field within the Fornax constellation in the southern sky, which the X-ray telescopes Chandra and XMM-Newton were already deeply observing. Additionally, this area was ideal for future deep observations in the infrared and submillimeter range. That fall, the observations began, and shortly after Christmas 2003, we were able to admire the first deep images (Plate 6).

The zoom into the HUDF in Plate 7 is astonishing because of its variety of colors and shapes. At the same time, it's important to know that the blue and the green in this picture correspond to the colors of visible light, whereas the red shows the invisible infrared information just outside the visible spectral range. The entire image contains only a single star from our Milky Way, which, in the top-left half of the picture, attracts attention with the spikes in its diffraction pattern. All of the other objects are distant galaxies with totally different shapes, sizes, and colors, but above all, they lie at different distances and are therefore different ages. The closer galaxies having nicely developed spiral or elliptical shapes emitted their light about five billion years ago when the universe was approximately nine billion years old ($z \sim 0.5$). The small blue dots in the background date back to a time about eight billion years ago when the universe was approximately six billion years old ($z \sim 1$), and the very faint red dots could possibly stem from the early days of the cosmos a mere 800 million years after the Big Bang ($z \sim 6.5$). Due to the expansion of the cosmos, their light now lies so far within the red region of the electromagnetic spectrum that they cannot be observed in any light but infrared. The new Hubble WFC3 camera's near-infrared capability has since detected galaxies even farther away. This more or less marks the limit of the observation possibilities using the current generation of telescopes. In the future, astronomers plan to look much further into the history of the cosmos; for example, employing the James Webb Space Telescope, NASA's next supertelescope in the near-infrared range, the Atacama Large Millimeter/submillimeter Array (ALMA), the gigantic interferometer consisting of sixty-four antennas for the millimeter and submillimeter range, or the next generation of 25–40 meter diameter ground-based telescopes.

In contrast to the fully developed spiral and elliptical galaxies in the near universe (as in Figure 1.2), the galaxies from the early cosmos are much more irregularly shaped. Some look like toothpicks; others look like bits of a bracelet. In some cases, several galaxies tangle together, closely interacting with each other. You can clearly see that these galaxies hail from a period in which major transformation processes took place. But one fact becomes particularly apparent when analyzing the pictures: these early galaxies were much smaller than they are today. This suggests that over the course of cosmic time the galaxies not only transform but also grow—especially through the fact that they "devour" one another.

As soon as we are capable of roughly determining the age of the galaxies, such as by measuring the redshift in their spectra, we can arrange the pictures of the galaxies according to their age. It's as if you were looking at a school yearbook showing various classes of students at different ages. Similarities exist between the photos, but you can see a clear development—for example, concerning clothes, hair length, or beards beginning to show on the boys. But above all, you notice a significant change in height from one year to the next. The students are growing—just like the galaxies.

As far as I'm concerned, galaxies belong to the "crowning glory of creation" in astronomy. They inhabit the densest regions of the universe and resemble pearls strung from the strings of the Cosmic Web. In the local universe, the galaxies present themselves in many different shapes and colors. In spiral galaxies like our own Milky Way, the stars in the outer regions are arranged like a flat disk, orbiting the center in elongated spiral arms together with gas clouds and dust clouds. The arms particularly stand out because of their young blue stars and ionized hydrogen clouds (see Figure 1.2), whereas the centers of the spiral galaxies often contain a thickening made up of old reddish stars, known as the "bulge." In elliptical galaxies, you basically see no gas or dust: only old stars with a reddish shimmer, which are arranged in a spherical or ellipsoidal structure. On closer examination, you notice that the elliptical galaxies share similarities with the bulges at the centers of spiral galaxies. Might it be possible that the ellipses and the spirals are merely two extreme versions of a family containing a considerably larger number of variants?

The well-known Sombrero Galaxy (Plates 8a, 8b) seems to support this idea. Here, a dust ring has arranged itself around a very well-defined, almost spherically shaped elliptical galaxy, which indeed shows signs of a galactic disk.[19] Thus, the Sombrero Galaxy is a mixture of a disk and an ellipse. Today, it is actually assumed that elliptical galaxies can turn into disk galaxies, and disks can turn into ellipticals. What plays an important role here are the merging and devouring processes that befall the galaxies during the course of their lives.

Matthias Steinmetz, my successor as director at the Astrophysical Institute Potsdam, was one of the first to manage making spiral galaxies and elliptical galaxies appear in a model ab initio ("from the beginning"—that is, from the early universe) and to monitor their development and transformation over the course of cosmic evolution. "We want to understand why some galaxies are spirals and others elliptical," Steinmetz said. "Where do these differences come from?" With regard to the big cosmological simulations described above, he explained: "Classical simulations show where the galaxies are located. But these simulations don't show what the galaxies look like in detail—and therefore whether they are spiral or elliptical galaxies."[20]

Galaxies, however, only account for a small fraction of the total mass of the universe. We can therefore regard them as "whitecaps" surfing on the ocean of hot gas and dark matter. In water, whitecaps form whenever the peaks and troughs of the waves grow so steep that the waves collapse. This is a chaotic, nonlinear process that is rather difficult to describe in physical terms. The evolution of the galaxies follows a similar process. It is such a complex phenomenon that so far, we only partly understand it. In addition to pure gravitational physics, which on the largest scale aptly describes the evolution of the cosmic structures, we also have to take into account the galaxies, the development of stars and black holes, the feedback of gas from the stars (for instance, through ultraviolet radiation), gas physics, stellar winds, supernova explosions and the effects of black holes, and possibly even magnetic fields. The galaxies are located within huge voids in the cosmos, but because their surrounding space influences their growth and maturation, simulating them requires an extreme level of numerical resolution and accuracy that even the newest and biggest supercomputers cannot handle.

In order to meet these requirements, even if only partially, Steinmetz and his team came up with a trick. First, they carried out a large-scale simulation of dark matter—similar to the Virgo calculations discussed above but not nearly as expensive. From the coarsely simulated volume, they cut out the area where they had identified a galaxy and traced it back in cosmic history to identify where the individual mass particles of the galaxy originated. Then they simulated this area of the virtual universe with its emerging galaxy once more, this time invoking a considerably larger number of particles as well as the entire necessary physics of baryonic gas, star formation, gas dynamics, etc. In addition, using the previous large-scale simulation, they considered all the important influences from outside the small-scale galaxy volume, including the gravitational and tidal fields of the large-scale cosmic structure. Using this method, they managed to reproduce not only the galaxies' details but their entire cosmic history. By simulating several different galaxies in this manner, they got a good impression of the various evolutionary stories in the galactic zoo. However, it should be noted that even given these elaborate simulations, the detailed physics of gas, star formation, dust, etc., can only be dealt with in a very rudimentary fashion. The galaxies simulated in Potsdam still do not come anywhere close to the beauty of the real galaxies.

Plate 9 shows sections from one of Steinmetz's simulations[21] in which a beautiful "grand-design" spiral galaxy is forming that can indeed be compared to the well-known galaxy NGC 1232 in Figure 1.2. The individual time steps show snapshots of the simulation, from perspectives above and from the side, at the respective redshift of the emitted light. The fresh primordial gas from the Big Bang is depicted in green; the first, youngest stars in light blue; the stars with ages between one and five billion years in yellow; and the oldest stars in red. These colors were chosen because the most massive stars usually radiate very brightly in the blue and ultraviolet range and only live for a very short time. Therefore, the blue areas in the galaxies show the young stellar populations. When the star population grows older as a whole, the bright blue stars die very quickly, leaving mostly the green, yellow, and orange stars. Our sun, for example, would be yellow in this image. In very old star populations, even the yellow stars have burned up, and only the weak but very long-lived red stars remain.

At a redshift of $z = 3$, which corresponds to a cosmic age of about 2.2 billion years, some young light-blue stars have, in fact, already formed, but most of the gas particles still find themselves in a virgin condition within the filaments of the Cosmic Web. At a redshift of $z = 2$ and a cosmic age of 3.3 billion years, many star clumps have already merged into a very irregular-looking blue protogalaxy that resembles a number of the faint dots in the Hubble Ultra-Deep Field in Plate 6. At $z = 1$, after about 5.3 billion years, most of the blue stars have vanished. In the meantime, after several smaller protogalaxies collide, a nice elliptical galaxy forms that looks more or less the same when viewed from above or from the side. Now, smaller galaxies and shreds of gas are continually being gathered up from the surrounding areas of the Cosmic Web. At a redshift of $z = 0.5$, about 8.6 billion years after the Big Bang, a small disk made up of young blue stars, fresh gas, and presumably dust has already lain itself around the original elliptical galaxy. This could more or less represent the evolutionary state of the

Sombrero Galaxy in Plates 8a and 8b. If this image were to describe the history of our own galaxy, our sun would only just have been created and would be visible as one of the blue stars in the outer regions of the disk. The galaxy is not always exactly in the middle of the simulated section due to the gravitational and tidal effects of the surrounding universe and its large-scale structure. Finally, if one waits for a redshift of $z = 0.1$ and therefore a universe age of 12.4 billion years, a beautiful spiral galaxy with a thin disk and well-defined arms evolves within the simulation. Inside the galaxy, a spherical thickening of old stars, related to the bulge of the spiral galaxy, reminds us of the galaxy's earlier elliptical phase. Just at this time, another smaller galaxy is swallowed within the simulation, exactly as actually occurs in the image of galaxy NGC 1232 (Figure 1.2). One can well imagine that this cycle will go on and on into perpetuity.

But the story is not over yet! Steinmetz has also simulated several cases in which large spiral galaxies merge with each other. During these events, the stars are whirled around to such an extent that elliptical galaxies form yet again. Such galaxy mergers will play a major role further on in this book. Our own Milky Way will share a similar fate in the foreseeable future. But we should take note of the fact that the images of the galaxies in the sky are only brief snapshots in an everlasting cosmic dance of transformation, of becoming, and of passing away. The evolution in the sky continues on Earth in the plant and animal kingdom!

Chapter Five

STAR FORMATION

THE MEGASTAR

Immediately after the decoupling of radiation and matter, the gas from the hydrogen and helium atoms reached a temperature of 3,000 K. From everywhere in the now transparent universe a hot, reddish-glowing wall of fire could be "seen" on the horizon. At that time, of course, there was nothing and nobody to witness anything. All of the complex structures of the cosmos had yet to be formed. In particular, not a single carbon atom or oxygen atom existed. Over a long period of time, the background radiation gradually cooled due to the expansion of the cosmos—a bit like a red-hot poker, which gradually cools down from red to brown until it merely emits heat. After about forty million years, the background radiation must have been more or less at room temperature, which actually could have been quite cozy. But with persistent inexorability, the cosmos expanded further and continued to grow colder and darker. This phase, which lasted approximately 200 to 300 million years, is called the "Dark Ages," a reference to Europe's barbaric Middle Ages (see Figure 2.3).

In the background, however, dark matter was carrying on with its mission, which had already begun in the first split seconds after the Big Bang, of massing and clumping together to form the Cosmic Web described in Chapter 4. Pregalactic objects—local accumulations of dark matter—grew due to gravitational instability from the primordial density fluctuations. Increasing amounts of matter streamed into these "bathtubs" of dark matter, causing the gravitational potential wells to continually deepen and leading the dark matter particles to move faster and faster within. While the universe as a whole was expanding and cooling, precisely the opposite was happening in the areas of the cosmos privileged with higher density. Dark matter was swarming together and in doing so, heating itself up. As we saw in Chapter 4, this process still continues today and will continue on into the future. The difference is that presently, structures the size of galaxy clusters and superclusters of about 10^{14} to 10^{15} solar masses are collapsing and heating up to temperatures of millions K, whereas in the days of the early universe, areas merely the size of a globular cluster (about 10^5 to 10^6 solar masses) were collapsing.

First, dark matter dragged normal baryonic gas, consisting of hydrogen atoms and helium atoms, into the denser areas, heating it up. Inside such a gas cloud, if the pressure due to the gravity rises higher than the pressure due to the temperature, the cloud can collapse and, in doing so, plant the seed for the first star. On the other hand, as we saw in the example of the air pump and the bicycle tire, we know that a gas cloud in the process of being compressed heats up considerably. In order to form a compact structure like a star, the gas cloud must be able to cool down efficiently. In the case of baryonic matter, this is made possible by the emission of radiation. Contrary to this, dark matter can never collapse and become a compact object.

An atom can produce radiation, for instance, if one of its electrons rises to a higher energy state. In the act of returning to the ground state, the electron releases a characteristic radiation that is precisely equivalent to the difference between the higher and the lower energy levels. This results in the previously mentioned characteristic Fraunhofer spectral lines. At temperatures of several hundred to perhaps 1,000 K, such as occurred during the phase of the universe we are dealing with at the moment, only heavy elements like carbon, nitrogen, oxygen, etc.—and also sodium and chlorine—were able to radiate well. This is well illustrated in the kitchen or in the laboratory by sprinkling a few grains of salt into the flame of a Bunsen burner or a gas stove. The practically invisible gas flame suddenly blazes yellow because the sodium atoms glow in characteristic yellow lines. In the early universe, however, only hydrogen and helium existed, as all the heavier elements had yet to be bred in the stomachs of the stars. In the case of hydrogen and helium, the electrons are so strongly bound to the atomic nuclei that they can only radiate properly above approximately 8,000 K. But such high temperatures didn't exist in the "Dark Ages" of the universe. Ergo, there is no radiation and therefore no cooling down and, consequently, no stars.

But once again, nature thought up a trick to overcome this hurdle. The fact is that at low temperatures atoms combine to form molecules. For example, within Earth's atmosphere oxygen does not occur in its atomic state, which would be chemically radical and harmful, but in its molecular state, O_2—always nice and neatly in twos. The same applies for nitrogen. Hydrogen combines to H_2 molecules at very icy temperatures beneath approximately 100 K. This is precisely what occurred in the universe about 200 to 300 million years after the Big Bang when all had sufficiently cooled down. The hydrogen molecules could suddenly perform a dance of rotations and vibrations, producing a whole zoo of molecular lines in the infrared range. Thanks to the hydrogen molecular lines, the gas cloud considered above was now able to emit its thermal energy very efficiently and thus, under the pressure of its own gravity, crash to the center of its mother halo of dark matter.

German astrophysicists Tom Abel and Volker Bromm, both working in the United States in friendly competition with each other, are pioneers in calculating the very first stars. They used the most modern supercomputers to bridge the vastly different orders of magnitude between the dimension of a pregalactic matter halo and the size of a single star.[1] Using three-dimensional simulations carried out early in the twenty-first century, Abel and Bromm both concluded that first, a molecular cloud of maybe one hundred thousand solar masses formed, which collapsed under its own gravity and in doing so, heaped up matter in its center to form a kind of baby star. This first star's interior reached such a high temperature and density that its nuclear fusion furnace began to burn. Hydrogen burned to helium, and shortly afterward, Fred Hoyle's magical carbon anomaly created the first carbon, along with nitrogen and oxygen—all of the elements necessary for the Bethe-Weizsäcker cycle (CNO cycle) of nuclear fusion mentioned in Chapter 3. With massive stars, this goes all the way up to iron. The baby star ignited when its central ball of gas attained roughly the mass of the sun, and as the core burned, further matter from the surroundings plummeted into it until it quickly reached the size of maybe one hundred or one thousand solar masses. All specialists agree that the first generation of stars, which evolved out of the primordial soup of virgin hydrogen and helium, must have had very large masses. In technical terms, these hypothetical objects are called population III stars, but so far, no such star has been discovered. Why is that the case? This star produced the very first light in the otherwise absolutely dark, cold universe. As we will see in more detail later, these massive stars mainly radiated in the blue and ultraviolet range of the spectrum. While doing so, they handled their energy in such a wasteful manner that they used their hydrogen supply up only about three million years after their appearance. Because these stars lived for such a short time, it is hardly possible today to see one still radiating.

We can imagine that such a star must have dramatically influenced its environment. The ultraviolet light rays were so high in energy that the hydrogen and helium atoms ionized once again. What is known as a "Strömgen sphere" of ionized gas wrapped itself around these stars. But even more importantly, the light of these stars was sufficient to break the bonds of the hydrogen molecules surrounding them. In this manner, a newly created star prevented any other gas cloud in its entire cosmic vicinity—not only in its own protogalaxy but far beyond—from producing stars, because it simply turned the cloud's cooling system off. Within its cosmic environment, it truly was the "megastar" and also the sole light source.

In addition, the megastar had another effect on its environment that can almost be described as suicidal. As we will see in more detail further on, massive stars ended their lives in a gigantic explosion: a supernova or even a hypernova. Due to the high pressure at their centers, the atoms of the heavier elements already bred there were welded together to form even heavier, partly radioactive elements, such as lead or uranium. A black hole might even have formed. Such objects could be the seed of the supermassive black holes that we see in the centers of galaxies today (see Chapter 8). The explosion cloud of such a star shot through the surrounding material at velocities of ten thousand or more kilometers per second and possibly swept a great deal of the gas out of its original protogalaxy. At the same time, the star "polluted" its environment with entire loads of heavy elements produced inside itself. However, only by creating this "environmental pollution" was life made possible for all the succeeding generations of stars, planets, and in the end, mankind. In the context of cyanobacteria, we will encounter a similarly life-creating environmental pollution on Earth in Chapter 9.

Megastars were created wherever dark matter in the early universe happened to have a particularly high-density contrast. Because the density fluctuations of dark matter initially increased linearly, the bathtub, which to begin with had the deepest gravitation potential, will also afterward always represent the deepest spot. In the cosmic regions where the megastars once emerged, we therefore today expect to find the centers of the biggest galaxy clusters. In our local universe, the deepest potential well of dark matter, to which our Milky Way is attracted, is likely located at the center of galaxy M87 in the middle of the Virgo Cluster, which contains a gargantuan black hole (Plate 23). It is quite possible that this bizarre object is the remains of the first megastar in our local universe.

THE STELLAR NURSERY

In our present-day universe, stars are created from dense gas clouds and molecular clouds. As opposed to the time of the early universe, our cosmos today contains large amounts of heavy elements that earlier star generations have bred. The gas masses, which collapse under their own gravity, no longer have difficulty getting rid of their surplus heat. At present, the exact process of star formation in the well-developed universe is still not completely understood. Unlike the very simple early universe containing only hydrogen and helium and no light at all, the active star-forming regions in today's universe encompass a chaos of hot and cold gases, molecular clouds with a complex chemical compositions, dust clouds, ultraviolet radiation, magnetic fields, and turbulence caused by supernova explosions. The beautiful images of the Eagle Nebula in Plate 10 (wide angle)[2] and Plate 11 (zoomed)[3] provide an idea of the complex and wild romances that take place in a star-forming region.

A comparison of Plate 10, taken with a small ground-based telescope, and Plate 11, taken by the Hubble Space Telescope, clearly shows that not only the Hubble can produce colorful pictures of the sky. The team responsible for public relations at the Space Telescope Science Institute in Baltimore has perfected the physically motivated coloring of astronomical images. What in Plate 11 looks like a winged fairy or a giant insect from a fantasy novel is in fact a gigantic tower of gas and dust in the Eagle Nebula, a delivery room for stars, in which the stars of a new open cluster are currently being born. The tower is about ten light-years in height, corresponding to roughly double the distance from the sun to the nearest star, Proxima Centauri. Some of the young stars have already formed and have hatched out of their cocoons of gas and dust. They can be recognized in the overview image in Plate 10. At the moment, their intensive ultraviolet radiation is just milling fascinating sculptures out of the dense, cold molecular clouds of the Eagle Nebula. Similar to how the wind in the desert blows away the light sand but leaves the heavier scree behind, some bizarre clumps of gas and dust resist being transported. Nevertheless, the starlight makes their surfaces glow in a ghost-like manner. At the end of the "elephant trunks" in the dense cocoons, some of which look like aeries, new baby stars are possibly being born right now. The density there has risen so high that due to their own gravity, individual gas and dust clouds detach themselves from the structure and collapse.

In the spring of 1774, German-British musician, mathematician, optician, and astronomer Sir Frederick William Herschel surveyed the sky with his huge self-built telescope while his sister Caroline meticulously noted the number of stars he found. One night he suddenly exclaimed to Caroline: "Truly there is a hole in the sky here." Herschel himself didn't attach too much importance to his discovery of the first dark nebula, but his sister looked into the whole matter more deeply and many years later published the first catalog of dark nebulae. Plate 12 shows the optical and infrared images of a Bok globule bearing the name Barnard 68.[4] In the 1940s, Dutch astronomer Bart Bok examined the dark nebulae in the Milky Way and named the particularly small and compact clouds "globules." These globules are presumably the remnants of the dense ends of elephant trunks within a big star-forming cloud similar to the one in the Eagle Nebula. European Southern Observatory (ESO) astronomer João Alves and his colleagues used the Very Large Telescope on Cerro Paranal and ESO's New Technology Telescope to analyze this globule in detail. Because of the dense dust, it is completely opaque in visible light, whereas in the infrared range, the stars behind it shine through. Interestingly enough, this globule is just on the brink of collapsing under its own gravity. Similar to a star, its gravitational pressure is only just balanced by the inner thermal pressure of the cloud. But as the cloud continues to cool down in the near future, at its center it will give birth to a young star.

Hence, the earliest phases of star formation occur in dense, opaque dust clouds and for this reason, are particularly hard to observe. At the beginning, the contents within these clouds are very cold: about 10 K, just above the temperature of the microwave background radiation. Only in the course of the advancing collapse do the gas and the dust in the center of the cloud gradually heat up and begin to radiate in the submillimeter range and in infrared light. For this long-wave radiation the cloud is transparent and the protostar can be observed. In the 1990s, a dramatic technological development resulted in very sensitive bolometer detectors for the submillimeter and millimeter range of the electromagnetic spectrum. Such detectors were used for the Submillimetre Common-User Bolometer Array (SCUBA) and, more recently, the SCUBA2 at the James Clerk Maxwell Telescope (JCMT) on Maunakea in Hawai'i. The thirty-meter telescope of the Institute for Radio Astronomy in the Millimeter Range (IRAM) on Pico Veleta in Andalusia also made use of this technology. Furthermore, from November 1995 until May 1998, the European Space Agency (ESA) operated the Infrared Space Observatory satellite (ISO), which, for the first time, carried a sixty-centimeter telescope cooled by liquid helium. The colder the objects are that the telescopes are supposed to observe, the more they themselves have to be cooled down; otherwise, their own heat can completely outshine the source signal. The scientists of the Max Planck Society, especially those from my former institute, made a major contribution to this development in the case of both the IRAM and the ISO. These new observation possibilities have also made it possible to measure the dust radiation in molecular clouds, which is just a little warmer than its environment, and to discover a great number of embryonic dust compressions and baby stars.

On August 25, 2003, NASA launched its Spitzer Space Telescope on a Delta rocket from Cape Canaveral in Florida. It carries a cooled infrared telescope with a mirror diameter of eighty-five centimeters and travels on an orbit around the sun, trailing behind Earth. Due to cleverly devised thermal shields that protect against the sun's and Earth's radiation, the telescope can be operated at very low temperatures, even without liquid helium. Along with the famous Hubble Space Telescope; the X-ray observatory Chandra; and the Compton Gamma Ray Observatory (CGRO), which now lies at the bottom of the ocean after a controlled crash, Spitzer is NASA's fourth and last "Great Observatory." Spitzer took the fascinating infrared image of the Sombrero Galaxy in Plate 8b, and it has also played a tremendous role in unveiling the earliest protostars.

Developed in a global collaboration between the United States, Europe, and Japan, the gigantic millimeter and submillimeter interferometer ALMA (Atacama Large Millimeter/submillimeter Array) boasts sixty-four twelve-meter antennas. Deployed on the five-thousand-meter Chajnantor plateau in Chile, it started full-scale operations in 2013. Its unique combination of angular resolution, spectral resolution, and sensitivity enables it to examine the dynamics of dust, gas masses, embryonic stars, and protostars in the star cradles of our Milky Way and galaxies even farther away. It will thus help further clarify the mechanisms of star formation.

Plate 13 shows a combined ALMA and visual image of the Herbig-Haro object HH 46/47, still entirely shrouded by a dark nebula.[5] Herbig-Haro objects are named after their discoverers George Herbig, one of my former colleagues at the Institute for Astronomy at the University of Hawai'i, who unfortunately died in 2013, and Mexican astronomer Guillermo Haro. They are bright, nebulous structures of gas and dust that, during a short phase of star formation, are ejected by their protostars while the protostars are still deeply embedded in their molecular clouds. In the visible light of the same region, we can only see a compact, opaque dark nebula—another Bok globule. It is located at a distance of more than one thousand light-years in the southern constellation Vela (Latin for the sails of the argonauts' ship in Greek mythology). The infrared and submillimeter images make this molecular cloud completely transparent and reveal a view of the baby star, almost like an ultrasound image of a child in the womb. Suddenly, we see an already brightly shining protostar emitting a fascinating bipolar outflow of gas. We will come across such "jets" in more detail further along in the book. We assume that all newborn stars go through similar development phases like this Herbig-Haro object. Our solar system must have been in a similar state about 4.5 billion years ago.

The Milky Way is believed to contain thousands of similar embryonic stars and protostars. The observation possibilities of these objects have once again improved dramatically over the last few years. In 2009, ESA launched its forth cornerstone, the Herschel Space Observatory, named after the famous astronomer. It carries a mirror with a diameter of 3.5 meters as well as cooled infrared detectors and can detect and analyze the thermal radiation of young stars with unprecedented sensitivity (see Plate 14).

The molecular cloud, massing together to form a star, first slowly rotates. As it collapses, it must get rid of its angular momentum. When ice skaters pull their arms in during a pirouette, they spin faster and faster. The collapsing gas cloud experiences something similar. The stronger it contracts, the faster it rotates. The question of how it eventually loses its angular momentum is currently a major topic of intensive research. A possible solution is the formation of binary stars. Approximately half of all stars are part of binary or multiple star systems. Another possibility is that a dust disk forms, from which a planetary system finally emerges. In our solar system, about 90 percent of the angular momentum is contained in the planets, whereas 90 percent of the mass resides in the sun. The planetary system develops out of a thin disk similar to the rings of Saturn but much denser. Such "accretion disks" arise in the cosmos whenever a rotating object collapses under its own gravity. In the equatorial plane above a certain angular velocity, the centrifugal force balances out gravity, whereas along the axis of rotation nothing counteracts gravity. Rotating gas clouds that collapse under their own gravity therefore automatically form into thin disks—not unlike a cook spinning a lump of dough into a pizza crust. We know such disks result from the accretion of matter onto a compact star remnant, such as a white dwarf, a neutron star, a black hole (see Chapter 8), or, on a gigantic scale, from the galactic disks of spiral galaxies.

The emerging star solves its angular momentum problem by "shedding its arms"; that is, it leaves the disk behind and can therefore continue collapsing toward its center. Numerical simulations show that within these disks, matter is transported to the inside and angular momentum is transported to the outside. Then, an initial gathering of only micrometer-sized dust particles is followed by the further accumulation of "fluff" and "dust bunnies," forming ever-larger matter clumps from which planets slowly emerge. This process has not yet been fully understood in theory. But in principle, we assume today that stars and planets develop in much the same way that Immanuel Kant and Pierre-Simon Laplace described in their eighteenth-century theses. In 1755, Kant postulated that the solar system evolved from a gaseous nebula and that in the course of this process, an aggregation of meteorites formed the planets. Laplace independently presented a very similar theory in 1796.

In the 1990s, observations strongly confirmed the theory that stars and planets form through accretion disks. Researchers measured the energy distributions in the light of young stars and noted a large amount of excess radiation from the infrared range right down to the millimeter range. This could only be explained by the emission of a dust disk. The most famous example of such a dust disk was discovered in the mid-1980s around the star Beta Pictoris. This star lies at a distance of sixty-three light-years in the Pictor constellation (the painter's easel) and is visible to the naked eye. It has about double the mass of the sun but is still very young. Since that time, as many as two dust disks have been discovered around this star in images from the Hubble Space Telescope, suggesting the existence of a planet roughly the size of Jupiter.

The Hubble telescope once again enabled a breakthrough in observing these dust disks. In 1995, English astronomer Mark McCaughrean, who at the time was at the Max Planck Institute for Astronomy in Heidelberg, Germany, and whom I hired shortly afterward to work at the Astrophysical Institute in Potsdam, very carefully examined the Hubble survey of the Orion Nebula with Bob O'Dell, then with Rice University in Houston, Texas. The Orion Nebula is about 1,500 light-years away from the sun and is one of the most active star-forming regions in our neighborhood. The two astronomers noticed dark specks in front of the backdrop of the bright nebula, which in some cases looked almost rectangular, like the shadow of a disk viewed from the side.[6] In all cases, these disks were considerably larger than our entire solar system. For the first time we were able to directly see the construction sites of planetary systems. Plate 15a shows the nicest examples of such "silhouette disks." In each case, you can see a baby star in the middle, about one million years old. Compared to the sun, which at an age of 4.5 billion years has just reached the zenith of its life, these stars have only just left the delivery room. They are still virtually lying in their mother's arms, newly born. These objects are called "protoplanetary disks" or, in the abbreviated form, "proplyds."

On the four pictures on Plate 15b we can see that young stars with protoplanetary disks live a fairly dangerous and uncomfortable life.[7] These close-up images from the Hubble telescope reveal structures resembling comets or tadpoles. These unusual shapes occur because the four fully grown massive stars forming the Trapezium at the center of the Orion Nebula blow the material around the proplyds away with their strong ultraviolet radiation and stellar winds. This produces a supersonic flow, called a "shock front," around the disks, which heats up the gas until it begins to shine. (This can be compared to the sonic boom of a jet fighter breaking the sound barrier.) Although the stars could, in fact, be producing planets in their disks, their hostile environment is already eroding them away. The process of planet formation must therefore happen very quickly. Scientists reckon that most of the protoplanetary disks in the Orion Nebula will not survive the next one hundred thousand years!

The object in the upper-left image of Plate 15b is especially fascinating. It shows the Orion proplyd HST-10. On the inside, you can see a green disk around the baby star. On closer examination, you get the impression that a stream of gas is being shot out at a vertical angle to the disk, which makes the shock front in the proplyd's direction of flight glow. Upstream, the jet spreads out even more. This phenomenon is similar to the bipolar outflow of the embryo star in Plate 13. Since the mid-1980s, in fact, we have known about the tightly bundled gas flows that young stars shoot out into the surrounding medium at supersonic speeds. The jet's axis is perpendicular to the protostellar disk, which suggests that the rotation of the star, or rather of the disk, is responsible for the jet.

We have knowledge of similar jets in accreting neutron stars or black holes, which we will examine in more detail in Chapter 6. The mechanism that produces these jets has yet to be completely understood, even today. But it is very likely that the magnetic fields generated by the disk's dynamo effect are responsible. This rotation winds these magnetic field lines up so tightly—similar to rolled-up spaghetti on a fork—that they form a kind of tower to bundle the strongly accelerated matter from the gas flow. The strength of this gas-stream bundling can be seen in Plate 16, which shows two young stellar objects at the top of elephant trunk towers in the star-forming Carina Nebula. This image, which reminds me of the Tower of Sauron from J. R. R. Tolkien's *The Lord of the Rings,* was taken in 2010 for the twentieth anniversary of the Hubble Space Telescope.[8]

These objects show a very complex structure with extremely finely collimated streams of gas shooting into the surrounding medium of the Carina Nebula at a velocity of nearly 250 kilometers per second. Taking a closer look, we see that the streams are not continuous but that single clumps are being fired out one after the other like machine-gun rounds. This suggests that the star goes through episodes in which the disk loses mass, and the jet is filled correspondingly. We also encounter this phenomenon in jets, considerably higher in energy, from rotating black holes (see Chapter 8). At the ends of the two jets that shoot out in opposite directions, you can see the bell-shaped, shining structures of the shock front that have developed wherever the supersonic flow powerfully hits the surrounding medium.

Chapter Six

WANDERERS IN THE SKY

THE THIRD DEFENESTRATION OF PRAGUE

Until August 2006, our solar system had nine planets: Mercury, Venus, Earth, Mars, Jupiter, Saturn, Uranus, Neptune, and Pluto. You could remember their order quite easily using the mnemonic: "My Very Educated Mother Just Served Us Nine Pizza-pies" or "Mom Victoriously Eats Mustard Jellybean Sandwiches Unlike No Person." The four inner terrestrial planets are stone orbs with a solid surface, like our Earth. They consist almost entirely of heavy elements. Under their solid crust is an inner shell of liquid magma, and at their centers is a fixed iron core. The next two planets are gas giants. Jupiter and Saturn, the largest planets in the solar system, are, like the sun, mostly composed of light elements with a small amount of heavier elements as well. The term "gas giant" is somewhat misleading because under the huge pressure of the overlying gas masses, the inner layers have liquefied or even solidified into hydrogen with metallic properties. However, they also may well possess a solid core of heavy elements. The next two planets, Uranus and Neptune, are ice giants. In contrast to the gas giants, they consist of only about 15 percent hydrogen and helium and for the most part are made of water ice, methane ice, and ammonia ice. Large ring systems with many moons, which share similarities to planetary systems, surround both gas giants and ice giants. Even further out lies the tiny planet Pluto—actually a double system of Pluto and its moon, Charon, orbiting each other.

112

On August 24, 2006, after two weeks of violent and passionate discussions, the General Assembly of the International Astronomical Union (IAU) in Prague, Czech Republic, surprised many by officially denying poor Pluto the status of a planet. Hence, the solar system now contains only eight planets. As compensation, Pluto was promoted to "king" of a new class of "dwarf planets," along with the asteroid Ceres and the object Eris (2003 UB313) discovered in 2003. On an Internet blog, an amateur astronomer indignantly complained that like on previous occasions in Prague, they should just throw the astronomers out the window: the Third Defenestration of Prague!

GREED AND GLORY

What had happened? To understand this, we must go back a little into the history of planet discoveries, some of which read like espionage thrillers. The original name "planet" was derived from the Greek language and means "the wanderer." Planets are "wandering stars" that move among the "fixed stars." In biblical times, seven planets were known, all of which are visible to the naked eye. Hence, the naming of the seven days of the week, which still refer to these celestial bodies in different languages: for example, Sunday and Monday to the sun and the moon; in Italian, Tuesday, Wednesday, Thursday, and Friday (Martedi, Mercoledi, Giovedi, and Venerdi) to Mars, Mercury, Jupiter, and Venus; and finally, Saturday to Saturn. The seven-day week is also related to astronomy through the lunar cycle of twenty-eight days: each phase of the moon—like a new moon, full moon, or half moon—spans seven days. In the ancient geocentric universe of Ptolemy and Aristotle, Earth was at the center of the cosmos, circled by all other celestial bodies. Therefore, the sun and the moon were correspondingly called "planets." In Bavarian parlance even today, people sometimes say: "The planet is burning pretty hot again!"

In 1543, astronomer Nicolaus Copernicus released his revolutionary work *De revolutionibus orbium coelestium,* which placed the sun at the center of the planetary system. The Catholic Church considered the heliocentric worldview to be heresy at that time, so Copernicus refrained from publishing his book until the year of his death. In the wake of the Copernican revolution, the sun and the moon lost their status as wandering stars while Earth was promoted to the status of a planet, and thus, there were only six left!

In early 1613, Galileo Galilei, who had discovered Jupiter's four large moons three years earlier and observed them meticulously thereafter, saw two faint dots of light in his telescope that had been absent from his previous recordings. Thinking they were fixed stars, he attached little importance to this finding, even when the distance between the two celestial bodies changed slightly over the following nights. Much later, it turned out that he had seen Neptune—and had missed discovering the first planet since ancient times and thus adding another diamond to his rich research crown. It was the aforementioned William Herschel who, in March 1781, serendipitously discovered the seventh planet of the solar system, named after the Greek god Uranus.

A calculation using the laws of planetary mechanics established by Isaac Newton and Johannes Kepler showed that Uranus moved in an orbit about twice the distance of Saturn's. But Uranus' path caused the astronomers headaches. In 1825, some irregularities were found, and in 1838, English Astronomer Royal and director of the Royal Observatory Greenwich, Sir George Biddell Airy, complained that according to his very accurate records, not only was Uranus significantly lagging behind its expected position, but even its distance from the sun showed anomalies.[1] The scientific world exploded in an uproar. Some suspected that another celestial body had thrown Uranus off course. Others believed that Newton's law of gravity might not apply so far out in the solar system or looked for a medium that might have slowed the planet. Still others, including mathematician Friedrich Wilhelm Bessel, suspected that another large planet with an orbit even farther away might be causing the perturbations. In 1843, the Royal Academy in Göttingen offered fifty ducats to anyone who could resurrect Kepler's laws within three years. This triggered a race against time. At the focus of the mysterious Uranus story was Sir Airy, who possessed the most accurate observational data but did not believe in the existence of a new planet. Rather, Airy thought that something might be incorrect regarding Newton's gravitational law. Both the young English mathematician and astronomer John Couch Adams and his French colleague Urbain Le Verrier independently attempted to calculate the celestial position of the suspected troublemaker.

Adams, who had obtained Airy's data in 1844, left a manuscript containing his results at the Astronomer Royal's home in September 1845 after trying in vain to personally meet with him. A few weeks later, Adams sent his corrected calculations, which now contained various possible solutions for a new planet. Airy, however, remained skeptical and asked Adams whether the new celestial body could also explain the anomalies in Uranus' orbit. Unfortunately, Adams never answered this very important question. Airy, in turn, did not encourage Adams to publish his calculations and they were never formally submitted to a journal. Adams also missed presenting his research to a meeting of the British Association for the Advancement of Science in Southampton, allegedly because he made a mistake in the date. The lack of communication between the high-ranking, very skeptical Astronomer Royal and the young, inexperienced mathematician is probably one of the keys to the obscure events that followed.

In September 1845, Airy visited the Paris Observatory and met its director, François Arago, as well as Le Verrier, who had published excellent work in theoretical astronomy. After discussions with Airy, Arago recommended that Le Verrier study Uranus' perturbations. In November 1845, Le Verrier presented an analysis to the French Academy of Sciences postulating the existence of a new planet. In June 1846, he published a second, improved calculation in the newspaper *Compte rendu*. Le Verrier tried in vain to inspire his director to organize a search for the new member of the solar system. Such an enterprise was extremely time consuming because the observer first had to accurately map out an entire section of the sky and then take a later look to determine whether any of the tiny dots of light had actually moved. Maybe the planet did not even exist?

But when Sir Airy received Le Verrier's latest publication at the end of June 1846, he found that its predictions almost exactly coincided with Adams'. In addition, Le Verrier's research confirmed that a new planet could explain the anomalies with the distance of Uranus. In July 1946, Airy, apparently convinced that a new planet existed, urged James Challis, director of the Cambridge Observatory, to search for it using the great Northumberland telescope.[2] At that time, Airy was the only person who knew of the two competing studies in England and in France, and he informed neither side. Toward the end of July, Challis began a relatively hesitant search for the new celestial body. As a starting point, he took the last coordinate Adams had specified and studied a large area of the sky several degrees long and one degree wide for several nights in July and August.[3] Later, it was discovered that he had actually observed the new planet on August 4 and August 12 but had not found it necessary to immediately compare the data from different nights. In a letter to Airy, he assumed that his search would probably continue well beyond the end of the year.[4]

On August 31, 1846, Le Verrier published his last and most accurate prediction regarding the new planet. His analysis was more detailed than Adams'; in particular, he predicted both coordinates of the
new planet while Adams had only calculated the length.[5] After futile
efforts to organize a search at the Paris Observatory, Le Verrier recalled
the young German astronomer Johann Gottfried Galle, who had previously sent Le Verrier his PhD thesis prepared at the Berlin Observatory. On September 18, Le Verrier wrote Galle a belated thanks and
begged him to look for the new planet at the predicted position: "Today
I want to ask the tireless observer that he devotes a few moments for
surveying a region of the sky, where there may be a new planet to discover. It is the theory of Uranus, which led me to this result." This
letter reached the Berlin Observatory on the evening of September 23,
1846. Actually, it would then have been the task of Director Johann
Franz Encke to get behind the telescope, but he had celebrated his fifty-
fifth birthday that very evening, therefore kindly leaving the nine-inch
telescope to Galle and his student Ludwig D'Arrest. This high-quality
refractor, prepared by Josef Fraunhofer, can still be admired today at
the Deutsches Museum in Munich. The eyepiece that Galle used, however, is on exhibit at the Babelsberg Observatory of the Leibniz Institute
for Astrophysics, Potsdam, where I had the pleasure to serve as director
from 1994 to 2001. At nightfall, Galle immediately began searching for
a disc-shaped object near Le Verrier's predicted position—originally

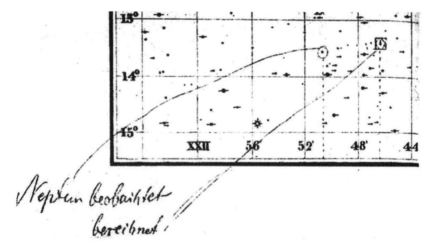

Figure 6.1. Section of the star map used by Galle and D'Arrest in the discovery of Neptune.
(Courtesy of the Leibniz Institute for Astrophysics Potsdam.)

without success. Then D'Arrest suggested they compare the sky with a highly accurate star chart recently published by Dr. C. Bremiker at the Prussian Academy of Sciences. No sooner said than done. Galle read out the coordinates of every star seen through the telescope while D'Arrest checked them off on the star chart. After only three-quarters of an hour, they had found the new planet. In Potsdam I could inspect the original star map myself, upon which Galle had noted in pencil: "Neptune observed—calculated" (see Figure 6.1).

When the two astronomers observed the object again the following night, it had moved considerably. Thus, any doubts about the discovery of a new planet were completely dispelled. The next morning, Galle wrote to Le Verrier: "Monsieur, the planet whose position you had sent me, *really exists*. On the same day that I received your letter, I have found an 8th mag star which is not recorded on the excellent star chart the XXI hour (created by Dr. Bremiker) from the collection of the Royal Academy of Sciences in Berlin."[6] This news spread around the world like wildfire. It was a triumph for the usage of mathematical methods in astronomy. For the first time, a celestial body had been discovered solely on the basis of theoretical predictions.

If one follows the events that happened in parallel in England, one can't help the impression that none of the participants placed too great an emphasis on the planet search. More intense, however, were the efforts to contest the fame of Le Verrier and Galle after their discovery. However, on October 14, Airy congratulated Le Verrier in a draft letter: "You are to (be) recognized beyond doubt as the real predictor of the planet's place," he wrote, then added a week later on the October 21:

> No person in England will dispute the completeness of your investigations, the sagacity of your remarks on the points it was important to observe, and the fairness of your moral convictions as to the accuracy and certainty of the results. With these things we have nothing which we can put in competition. My acknowledgement of this will never be wanting; nor, I am confident, will that of any other Englishman who really knows the history of the matter.

In truth, however, the Englishmen had already knitted a legend, according to which John Couch Adams and John Challis were entitled to at least as much fame in the discovery as Le Verrier and Galle. Adams' original prediction of the planet's position was only one degree removed from Le Verrier's, and Challis had spotted the planet even before Galle. Of particular concern was the allegation that the astronomers in Cambridge would have achieved success just as quickly as the Germans if they had gained access to Berlin's star map. The highlight of this nationalistically colored dispute culminated in the English claim to name the eighth planet—Neptune almost ended up as "Oceanus." Ultimately, Le Verrier was unable to counteract the undocumented assertions, published in several letters from Airy to the journal *Astronomische nachrichten,* and therefore had to willy-nilly share the glory with the English astronomers. However, the international astronomical community at least compromised by giving the new planet Le Verrier's proposed name, "Neptune," and the Royal Society awarded Le Verrier the Copley Medal in 1846. Le Verrier himself never properly recovered from the drama. Although he was later appointed director of the Paris Observatory, he was considered a bitter, unjust man, whose employees rose against him to force his (temporary) resignation.

The original works from England, in particular Airy's extensive correspondence, were not accessible for over a century. As of the 1960s, they were considered lost—possibly stolen. Only in 1998, after the death of astronomer Olin Eggen, who had worked in the 1960s in Greenwich, were Airy's "Neptune Files" discovered in Eggen's estate in Chile next to several meters of very valuable books from the Greenwich archives. Their analysis revealed detailed evidence of cheating by Greenwich and Cambridge. As it turned out, Adams' calculations had been too inaccurate to lead to the discovery of the new planet. Also, Challis had indeed possessed one of the Berlin star maps, on which the appropriate sky region had been recorded. He would have discovered Neptune in August 1846 if he had only made the comparison. In a bitter exchange of letters, Airy was subsequently attacked because he had not forwarded Le Verrier's calculations to the observatory in Cambridge in order to secure the planet discovery for England. Following Nick Kollerstrom's accurate evaluation of these files,[7] the discovery of Neptune must be attributed to Le Verrier and Galle alone.

THE KING OF DWARFS

We should keep this history in mind as we now turn our attention to the fate of Pluto. As early as September 30, 1846, one week after the discovery of Neptune, Le Verrier had expressed the presumption that there could still be another unknown planet outside of the Neptune orbit. He was by no means the only one who believed in another large celestial body, and thus another dogfight for the fame and glory of discovering a new planet began. The best-elaborated predictions for the position of a hypothetical trans-Neptunian planet came from two US astronomers who were fierce competitors: William Henry Pickering in 1909 and Percival Lowell in 1915. Both dealt with the same problem, using different methods, and both arrived at quite different predictions. Lowell, from one of the wealthiest families in New England, was an avid amateur astronomer. Giovanni Schiaparelli's discovery of the "canals of Mars" inspired Lowell to theorize that these could be artificial irrigation channels. The idea of "little green men" on Mars fell on fertile ground with the sensationalist media and science fiction authors and partially shaped the image of the red planet we know today. To promote his hobby, Lowell and others founded the Flagstaff Observatory in Arizona in 1894, which was later renamed the Lowell Observatory in his honor. It still provides important contributions to astronomy.

Until the end of his life, Lowell remained deeply absorbed in searching for this predicted "Planet X"—but in vain. However, other US astronomers remained obsessed with the idea. Both Milton Humason and Vesto Slipher, whom we have already met in connection with Hubble's discovery of the expanding universe, became involved. In 1929, Slipher, director of the Lowell Observatory, employed young, enthusiastic Clyde W. Tombaugh, the son of a farmer, as a night assistant and entrusted him with the task of searching the sky for the trans-Neptunian planet using photographic plates. Tombaugh systematically worked through the zodiac signs of Pisces to the constellation Gemini, always comparing two consecutive exposures with a blink comparator. In such a device, which belonged to the default instrumentation of each observatory before the introduction of digital detectors, two plates of the same sky field are placed under a microscope while the light is quickly switched back and forth with a tilting mirror. This comparison method can detect a motion in the sky for even the tiniest specks of light. However, scanning a complete pair of plates takes about three days. On February 18, 1930, after he had searched hundreds of plate pairs and millions of stars, Tombaugh finally located a tiny object that had moved about one-tenth of a degree across the sky within ten days. Due to a monumental commitment of financial and human resources—and nearly two generations after the original prediction—a trans-Neptunian planet had been found at last!

Although the astronomers were originally somewhat cautious, the discovery of a new planet spread throughout the world press in no time. Harvard Observatory director Harlow Shapley, who had participated in the "Great Debate" about the galaxies a few years earlier and, in a sort of second "Copernican revolution" had moved the sun from the center of the universe to the outskirts of our galaxy, described the discovery as "the most important since the discovery of Neptune" in an article in the *Times*.

Of course, every child needs a name, and the right to propose the naming of a new solar system object lies with the discoverer. It should be noted, however, that planets are named after figures in Greek or Roman mythology. All sorts of names were proposed for the new planet, and the press, especially, was actively involved. The name Pluto was finally proposed by eleven-year-old schoolgirl Venetia Burney of Oxford, England. Interestingly, her uncle had already proposed the names of the Mars moons, Phobos and Deimos. Pluto is the ruler of the underworld in classical mythology. Venetia knew this but was also well aware that the name starts with Percival Lowell's initials, which would particularly reward his lifelong search. In the same year, Walt Disney's favorite dog, named after the planet, made his debut.

Both the observatory in Flagstaff and the international community of astronomers enthusiastically accepted the name, and PL became its official abbreviation. However, it soon became evident that Pluto was not likely to be the "Planet X" postulated by Lowell and many others. Its orbital inclination, as calculated by employees of the Babelsberg Observatory, turned out to be unusually large and more akin to an asteroid. Its orbital period and the mass estimated from its brightness were, by far, insufficient to explain the presumed gravitational disturbances in the orbit of Neptune. These later would turn out to be measurement errors.

Pluto's moon, Charon, was not discovered until 1978, again at the Lowell Observatory. James W. Christy named the moon after the mythological figure of the ferryman who carries the souls of the dead across the River Styx to Pluto's underworld, Hades. However, Christy may have had personal motives as well because his wife, Charlene, was nicknamed "Char." From the orbit of the two celestial bodies, Pluto's mass was determined to be only about 1/400 that of Earth's, with a size comparable to hundreds of other minor bodies situated in the asteroid belt between Mars and Jupiter and in the Kuiper Belt beyond Neptune, the "construction rubble of the solar system."

Tombaugh himself must soon have become aware of this fact because he continued to search for the "needle in a haystack" for another thirteen years. During that time, he inspected about thirty million stars and three-quarters of the entire sky. He discovered a plethora of new astronomical objects but not a single new planet. What at first appeared to be another triumph of celestial mechanics was, ultimately, only a serendipitous discovery due to his diligence and perseverance.

The end of Pluto's status as an "ordinary" planet was ushered in by the work of US astronomers Dave Jewitt and Jane Luu, who discovered trans-Neptunian bodies beyond Pluto from Hawaii in 1992 and gave the "Kuiper Belt" its name,[8] as well as Michael Brown, who found several large Kuiper Belt objects from 2002 to 2005. Today, the technique of monitoring the sky is so far advanced that computers elegantly and quickly do the same work that Tombaugh had to laboriously perform with his naked eyes. As a result, much weaker objects can be found. On January 5, 2005, Brown and his colleagues identified the object 2003 UB313 in the Kuiper Belt. To confirm the candidates for planetary status, extensive follow-up observations must be carried out. However, Spanish astronomers obtained the pointing coordinates of one the telescopes used in the search from a public website and published the discovery of the minor planet Haumea that Brown's group had identified a year earlier. To secure his priority, Brown was forced to announce the discovery of 2003 UB_{313}, only nineteen hours after the Spanish claim, along with another object in the Kuiper Belt. He gave this object the nickname "Xena" after the heroine of the popular fantasy series. Later, the same group discovered a moon around 2003 UB_{313}, which they gave the nickname "Gabrielle," also a character from the same television series. Using detailed radio measurements with the Institute for Radio Astronomy in the Millimeter Range (IRAM) telescope in southern Spain, my colleague Frank Bertoldi from the Max Planck Institute for Radio Astronomy in Bonn successfully determined the diameter of Xena, which turned out to be slightly larger than Pluto.[9] NASA and a number of newspapers and television stations now claimed Xena to be the tenth planet of the solar system and demanded a proper authorization from the IAU.

Every three years, IAU members officially meet in a different city around the globe. Usually 2,000 to 2,500 out of nearly 10,000 members attend the two-week event. In addition to various specialized scientific events, the plenary sessions, in which members vote on essential resolutions, are particularly important. For example, members might vote on the definition of coordinate systems or the exact time of the coming new year, with the necessary leap seconds, on New Year's Eve. In 2006, the twentieth IAU General Assembly was held in Prague, where Tycho Brahe and Johannes Kepler had written the history of astronomy nearly four hundred years before. I had the honor of traveling to Prague as the national representative of the German astronomers. Before the meeting, rumors abounded that an important decision about the planets was to be made. In the solemn opening ceremony, the IAU president said he hoped this decision would be remembered in history as the "Prague Declaration." Afterward, the whole affair turned out to be a big public relations nightmare due mostly to the IAU's suboptimal management. Some years earlier, a nineteen-member IAU commission had been concerned with the exact physical definition of a planet. This commission broke up in a quarrel and put forward two mutually opposing recommendations. In one variant, a planet should be defined "from the inside"—that is, as having so much mass that its own gravity liquefies the rocks in its interior to form a hydrostatic equilibrium and an essentially round shape. Usually, this is the case for rocks more than eight hundred kilometers in diameter. By this definition, Pluto was a planet but so were many other minor bodies in the asteroid belt and the Kuiper Belt.

The other group argued that the dynamic effects of a planet on its neighborhood in the solar system must be taken into account: only a body that possesses enough gravity to dominate its entire surroundings within the solar system and thus has cleared its environment of smaller bodies should be called a planet. According to this definition, Pluto was no longer a planet because it was a member of the Kuiper Belt. After this original commission proved unable to agree, the IAU president appointed a seven-member commission that contained social scientists and education specialists in addition to astronomers and planetary scientists. They were well aware of the dramatic impact a new planet definition had on textbooks and encyclopedias as well as the general public. This commission worked in Prague in absolute secrecy. Then, the IAU executive made the serious mistake, in my opinion, of informing the world press about the new commission's draft resolution before any other astronomer saw it. Likely, many astronomers shared the same feelings I had at the time. I was interviewed by a journalist from the leading German weekly, *Der Spiegel,* regarding my opinion on the draft resolution, and I had to ask him to read it to me prior to commenting. According to this resolution, the "hydrodynamicists" had prevailed. Every celestial body that was heavy enough to be round, therefore, would be a planet as long as it did not orbit a planet. In addition to Pluto and Charon, the asteroid Ceres—which had been documented for over two hundred years—as well as the newly discovered 2003 UB_{313} (Xena) would qualify for planetary status. A further commission still needed to investigate the roundness of an additional twelve planetary candidates, including the asteroid Vesta, which was nicely round but badly "dented" by a collision not too much earlier. It was not entirely clear how deep a crater could actually be before a body was no longer considered round. For many of the candidates on the planet waiting list, there were simply too few accurate observations. In addition, one could easily imagine that the sensitive digital sky surveys would discover up to about forty more planetary candidates in the future.

As a national representative, I initially requested more consideration time to ask my colleagues in Germany for their opinion by e-mail. Such feedback came promptly and numerously. Most believed that this draft resolution should be rejected. Apart from the fact that the new planet definition failed to clarify the unresolved issue about the upper mass limit in extrasolar planetary systems and therefore whether a brown dwarf would be a planet or a star, my colleagues disliked increasing the number of planets and the concomitant devaluation of the planet definition. Most colleagues favored keeping the nine historical planets or, if necessary, dispensing with Pluto as a planet.

The next day, almost all major newspapers reported that astronomers in Prague planned to announce an addition to the solar system. The first discussion of the resolution during a plenary session in the second week of the conference turned into a fiasco. Immediately after the IAU president's inaugural address and the seven-member commission chairman's explanation of the resolution, IAU members formed long queues on both sides of the aisle in front of the hall microphones to comment. Their tenor was almost entirely negative. The president tried to stifle the discussion in a somewhat undiplomatic way, but in the end, he had no choice but to let the concentrated expressions of discontent run free. In the discussion, transmitted live on the Internet, I got the last word by chance and was able to again reaffirm the predominantly negative attitude of the German astronomers. A straw poll took place in which about 80 percent of those present opposed the resolution. The much-heralded plan had resoundingly failed.

Within hours, the IAU swung around and decided to include the ideas of the dynamists in the planet definition. To counter the criticism of the discoverers of extrasolar planets, the new definition was explicitly limited to the solar system. A straw vote called on short notice showed overwhelming support for the new draft resolution. The discussion that ensued and the vote at the plenary meeting mainly revolved around the fate of Pluto. The members decided to change it, as well as the other round minor bodies in the solar system, to the status of a "dwarf planet," which must always be placed within quotation marks to indicate that it is not a real planet. As compensation it was allowed to keep Charon as a moon and was also defined as the prototype of a new class of trans-Neptunian "dwarf planets" whose name (pluton, plutinids, plutonian objects?), in the end, unfortunately could not be agreed upon.

On August 24, 2006, the General Assembly gave astronomers a public attention hitherto never reached. Some twenty international television crews and countless journalists reported live from the event. Shortly after the vote, CNN headlined "Pluto demoted," and all the newspapers wondered at the dramatic shift in opinion among astronomers. I was particularly impressed—as on other occasions—by the Internet encyclopedia Wikipedia, which correctly stated the new definition of planets and "dwarf planets" only twenty minutes after the vote. After defining the term "dwarf planet," the IAU immediately named the largest "dwarf planet," 2003 UB_{313} (formerly Xena), Eris. This is the name of the Greek goddess of strife and discord, a carefully penned label for the celestial body that set the discussion about Pluto and the rest of the planets in motion. As a kind of revenge, the American Dialect Society chose the word "plutoed," as well as the verb "to pluto," as the word of the year for 2006. Since that time, it has referred to a dramatic degradation or devaluation of a thing or a person in the American language.

Despite the awful public relations, the Pluto story has one positive effect, in my opinion: it shows the progress of scientific knowledge, sometimes going in serpentine directions and yet often at a breathtaking pace. It also shows how new insights into the nature of the cosmos are directly connected with our self-understanding and our own whence and whither. In several interviews, I was asked how on Earth school kids should memorize the names of the planets in the future. It's quite simple: "My Very Educated Mother Just Served Us Nuts."

COUNTLESS WORLDS

A few centuries ago, Dominican monk Giordano Bruno postulated that the universe contains an infinite number of planets and countless numbers of creatures inhabiting other planets. On February 17, 1600, he was burned at the stake for his heresy in the Campo di Fiori in Rome and has never been fully exonerated despite the Church's now much more liberal views.[10] Meanwhile, astronomers had already discovered a huge number of planets outside the solar system. The search for signs of life on extrasolar planets is one of the most fascinating research goals—a kind of Holy Grail for astronomers. Planets around other stars are extremely hard to identify because when viewed from a long distance, their feeble light is completely swamped by the glare of their parent star, making it almost impossible to obtain their direct images. However, a number of techniques can aid in this process, most of which are based on the effects the planet has on its central star.

The first hopes of discovering extrasolar planets came in the 1980s when astronomers using infrared observations found disks of dust surrounding otherwise quite normal stars (see Chapter 5). Radio astronomer Alexander Wolszczan announced the first true exoplanet discovery in 1994 after finding a number of planet-sized objects orbiting a pulsar, a spinning neutron star, in the Virgo constellation. (We will learn more about neutron stars in Chapter 7.) This discovery was made possible by observing small, regular variations caused by the gravitational effects of the planet on the otherwise extremely stable radio-timing signal of the rapidly rotating neutron star. Although the astronomers accepted these findings, many were disappointed because the unfriendly environment around a rapidly spinning neutron star was not the place to look for habitable planets.

The breakthrough came in October 1995 when two astronomers from the University of Geneva, Michel Mayor and Didier Queloz, using the relatively small telescopes of the Observatoire de Haute Provence, published the first definitive discovery of an extrasolar planet around the normal star 51 Pegasi. A few months later, US astronomers Geoffrey Marcy and R. Paul Butler, using the Lick and the Keck telescopes, confirmed the discovery and presented two more extrasolar planets. Both teams used the "Doppler-wobble" method. Through high-precision spectroscopy, this method measures the central star's extremely small motion due to its orbiting planets. Jupiter, for example, only approximately orbits around the sun. According to Kepler's

laws, both celestial bodies actually revolve around their common center of gravity. As a result, the sun periodically dances around the common center of mass, which is within the solar surface. Measuring the
extremely small radial velocity of the star, which corresponds to walking speeds of a few meters per second, requires an extremely precise
speed trap. Like the radar speed traps that police use, this works with
the help of the Doppler effect; that is, the shift of the spectral lines in
the spectrum of the star. Using Kepler's laws, this Doppler-wobble, or
radial-velocity, method can determine the mass of the planet, know-
ing its orbital period.

The following decade saw a surge of new discoveries in this field,
partially driven by the construction of new, dedicated ground-based
instruments and advanced technology such as adaptive optics that
allow the direct imaging of some planets in special configurations,
better spectrographs, better detectors, and more powerful computer
algorithms. By August 2006, the *Extrasolar Planets Encyclopedia*
listed exactly two hundred known planets outside our solar system.
By April 3, 2015, this list had skyrocketed to 1,906 planets in 1,202
planetary systems, of which 480 are multiple-planet systems.[11]

The major growth of exoplanet discoveries in the last decade is
due to the introduction of a new indirect detection scheme: the planet
transit method. If we are lucky, we view the plane in which the planet
orbits its host star from the edge, so the planet regularly transits in
front of its star. On June 5, 2012, in an event organized by the Institute for Astronomy, Hawai'i (IfA), a large crowd of spectators assembled on Waikiki Beach in Honolulu to watch Venus transit in front of
the sun, a spectacle that only occurs roughly every 115 years in pairs
separated by 8 years. It was fascinating to watch—along with fifteen
thousand spectators—the little "beauty dot" of Venus pass over the
whole face of the sun during this six-and-a-half-hour event. The transit of an exoplanet is a similar but very rare phenomenon because the
distant planetary system must be viewed almost exactly from the edge.
In the case of an Earth-like planet around a solar-type star, the likelihood of catching a transit is only 0.5 percent. However, the probability
rises to about 10 percent when viewing a Jupiter-sized planet close to its
star. When the planet passes in front of its host star, it occults a minute
fraction of light, reducing the star's brightness. With modern detectors and a very stable observing platform, you can measure this tiny
abatement, and if you observe long enough, you should see the signal
repeating. In 1999, Mayor's group announced the first discovery of

an extrasolar planet's transit, utilizing a small ten-centimeter telescope with a modern CCD (Charge Coupled Device) detector. The planet HD 209458b was originally discovered with the Doppler-wobble method, but the observed transit signals confirmed the existence of the planet beyond any doubt and opened a new avenue for discovering extrasolar planets.

The breakthrough in the field of transit discoveries came with dedicated space observatories. The French space agency Centre National d'Études Spatiales (CNES) launched the Convection Rotation et Transits planétaires (COROT) satellite in 2006, the first mission specifically designed to find and study exoplanets. With its twenty-seven-centimeter telescope, it has indeed discovered the first extrasolar planets using the microscopic eclipses of transiting planets. In 2009, NASA launched its first extrasolar planet mission, Kepler. The satellite carried a single-purpose instrument with a 0.95-meter telescope and a focal plane of forty-two CCDs, covering a huge field of view of about twelve degrees in diameter. For more than four years, it continuously stared at a sky field in the Cygnus constellation, where it simultaneously measured the brightness of more than 150,000 stars, recording extremely accurate photometric measurements every thirty minutes. In May 2013, the Kepler spacecraft unfortunately lost the second of its four reaction wheels and therefore could not continue its nominal mission. NASA scientists and engineers, together with the spacecraft contractor, have since designed a fancy new emergency operation mission, K2, which has enabled it to continue the exoplanet observations. NASA has also selected a second Explorer-class exoplanet craft, the Transient Exoplanet Survey Satellite (TESS), to be launched in 2017, which will be able to find Earth-like planets closer to Earth. One of our IfA faculty members, Andrew Howard, is a co-investigator in this mission.

The statistics gained from exoplanet discoveries have yielded several surprises. The radial velocity method is the most sensitive for large, Jupiter-sized planets rather close to their host stars. Indeed, Jupiter-size and larger objects dominated the statistics of the two hundred known exoplanets in 2006. Surprisingly, the typical geometry of other planetary systems is completely different from our own solar system, with large planets that are much closer to their host stars. The most extreme cases are the "hot Jupiters"—large gas planets that orbit closer to their star than Mercury does in our solar system. With such a hot Jupiter, the side facing its star has a temperature comparable to the stellar surface while the side facing away is much colder—causing a devilish climate. Although selection effects obviously dominate the statistics, the fact that large planets are discovered so close to their host stars has triggered a lot of new theoretical work regarding the evolution of planetary systems, including our own. Indeed, the fact that Jupiter must have wandered from the solar system's outer regions close to Earth's orbit is now regarded to be a major factor in the solar system's formation and evolution. In this way, water could have been brought to Earth's orbit from the cold, far reaches of the solar system.

The chemical analysis of exoplanet host stars yields other important information: the Jupiter-like gas giants apparently exist only around stars that, like the sun, contain large amounts of heavy elements in their atmospheres. Giant planets have not been discovered around chemically young stars, which formed from almost virgin gas after the Big Bang with very few heavy elements. This suggests that the formation of gas giants requires large amounts of heavy elements. On the other hand, smaller planets seem to occur around host stars with a wider range of metallicities.[12] The heavy elements in the prestellar cloud produce dust grains in the disk around the baby star that gradually stick together to form larger and larger bodies—from "fluff" and "dust bunnies" to rocks, asteroids, and comets—until the forming body's gravity ultimately becomes sufficient to slowly pick up the surrounding disk material to form large planets.

Given the important role of Jupiter and Saturn, it was therefore no coincidence that the solar system formed relatively late in the development of the cosmos, about nine billion years after the Big Bang. Before that time there were too few heavy elements present. It also comes as no surprise that the sun is located in the intermediate regions of a spiral galaxy. The central region of the Milky Way, in particular the bulge, was created by old red stars that emerged very early in the history of the cosmos and therefore lacks heavy elements. The young stars on the very edge of the spiral galaxy formed from almost pristine gas recently accreted from primordial Big Bang matter. For these stars, the chances for giant planets remain low due to the spare amount of heavy elements. Our sun is therefore located exactly in the right place in space and time to form the solar system!

Exoplanet Discoveries in Hawai'i

In recent years, the University of Hawai'i at Mānoa, particularly the IfA, has grown into arguably one of the world's most active centers for excellence in the discovery and the study of extrasolar planets. As director I have had the pleasure and honor to accompany this development. A very active, young group of faculty members, postdoctoral researchers, and graduate students are carrying the field forward, and I would like to feature some of their recent discoveries.

Adam Kraus, a postdoctoral researcher at IfA, together with his colleague Michael Ireland, have captured the first direct image of a "baby planet" in the process of forming around its star. The object LkCa 15 b looks like a hot "protoplanet" surrounded by a swath of cooler dust and gas that is falling into the still-forming planet. At the age of about two million years, not much older than the Hawai'ian island of Oahu, the youngest planet ever discovered sits inside a wide gap between its young parent star and an outer disk of dust (see Plate 17). The two astronomers combined the power of one of the ten-meter Keck telescopes' adaptive optics system, which cancels out the air turbulence above Maunakea, with a classical technique called "aperture mask interferometry." This involves placing a small mask with several holes in the path of the light collected and concentrated by the giant telescope. Using this method, Kraus and Ireland can achieve the sharpest contrast ever, allowing them to resolve the protoplanetary disks around stars and see gaps in the dust where developing planets may be hiding.[13]

Using planetary transit data from the NASA Kepler spacecraft, another team of astronomers, including Nader Haghighipour from the IfA, has discovered the first system in which two planets orbit around two stars—a Double-Double. This system, called Kepler-47, is similar to the iconic planet Tatooine, the home of Luke Skywalker, in the *Star Wars* universe. This discovery demonstrates that complete planetary systems can exist around a binary star.[14]

An international team of astronomers, including Michael Liu and Eugene Magnier from the IfA, has discovered an exotic free-floating young planet that does not orbit a star. The object, called PSO J318.5–22, is just eighty light-years away from Earth and has a mass only six times that of Jupiter and an age of only twelve million years. Using the Panoramic Survey Telescope and Rapid Response System (Pan-STARRS) PS1 telescope, they identified this unique object from its faint and strangely colored light and then employed other telescopes to show its similarity to gas-giant planets orbiting young stars. But the "lonely planet" PSO J318.5–22 is all by itself, without a host star. How such a planet could form is still an unsolved question. Based on its distance and its motion through space, the team concluded that PSO

J318.5–22 belongs to a collection of young stars called the Beta Pictoris moving group that formed about twelve million years ago, including the famous star Beta Pictoris itself (see Chapter 5).[15]

A team of astronomers led by the IfA's Andrew Howard has detected the first Earth-sized planet outside the solar system with a rocky composition like that of Earth.[16] This exoplanet, known as Kepler-78b, orbits its star very closely every eight-and-a-half hours, making it much too hot to support life. After obtaining the Kepler spacecraft's telltale planetary transit data, which determine the size of the planet, they were able to measure its mass through high-resolution spectroscopy at the Keck Observatory on Maunakea. A handful of planets with the size or mass of Earth have been previously discovered, but this is the first in which both the mass and the size are known, allowing astronomers to measure its density and confirm that, like Earth, the planet is primarily made of rock and iron. An independent study by the Geneva Observatory in Switzerland confirmed these findings. The host star of Kepler-78b, somewhat smaller and less massive than the sun, is about four hundred light-years away. Anyway, it would probably not be worthwhile to try to travel there because the temperature on this planet must be hellishly hot, and the rock surface most likely consists of molten lava (see Plate 18).

Although the Kepler spacecraft is now crippled, it has completed its main mission objective to determine what fraction of the stars in our galaxy have potentially habitable planets. Exoplanet pioneer Geoffrey Marcy from the University of California, Berkeley; Andrew Howard; and UC Berkeley graduate student Eric Petigura, who visited the IfA for a year, analyzed all the Kepler discoveries and came to the fundamental conclusion that about one in five stars like the sun have planets about Earth's size and a temperature that allows for liquid water on the surface—thought to be a precondition for life. This amazing finding means that "when you look up at the thousands of stars in the night sky, the nearest sun-like star with an Earth-size planet in its habitable zone is probably only 12 light years away and can be seen with the naked eye."[17] In 2014, Petigura, Howard, and Marcy published these findings in the *Proceedings of the National Academy of Sciences* and received the academy's prestigious Cozzarelli Prize.

Chapter Seven

THE STELLAR CEMETERY

RED GIANTS AND WHITE DWARFS

Stars are giant balls of gas kept in equilibrium by their gravity, on one hand, and by their interior gas and radiation pressure, on the other. This pressure, produced by the star's interior heat, is ultimately radiated away from its surface. In the nineteenth century, William Thomson (Lord Kelvin) and Hermann von Helmholtz pointed out that a star like the sun (but without additional internal energy sources) could radiate in this manner for about twenty to thirty million years as it slowly contracted. Lord Kelvin also derived a minimum age for Earth that far exceeded all previous estimates. Today, we know that a star undergoes this Kelvin-Helmholtz contraction during the early stage of forming from the embryonic gas cloud. However, this cannot be the mechanism from which the sun draws its energy today. Only at the beginning of the twentieth century did it become possible to determine Earth's exact age using radioactive methods. It has existed for approximately 4.6 billion years—well over one hundred times higher than the Kelvin-Helmholtz time. This presented astronomers with significant challenges because the sun could not possibly be much younger than Earth. In 1920, English astronomer Sir Arthur Eddington was the first to propose that the stars derive energy from the nuclear fusion at their centers. In this process, hydrogen atoms fuse to form helium, which releases enough energy for the sun and the other stars to radiate for billions of years. For more than fifty years, researchers have tried to bring this solar fire down to Earth to produce energy in fusion reactors and are making progress in solving mankind's energy problems, hopefully in this century.

Rudolf Kippenhahn, one of the pioneers of stellar evolution theory, described the story of stellar evolution in his book *100 Billion Suns: The Birth, Life and Death of the Stars* in a beautifully vivid fashion.[1] With his brilliant and easy-to-understand lectures, Kippenhahn was a teacher who helped me and many others find our passion for astrophysics. Although his book has more than a third of a century under its belt, and many outstanding issues of its time have now been resolved, it provides an excellent reference and overview. In principle, the sun would have enough supply of hydrogen to last for nearly one hundred billion years. However, only a small part of the sun participates in nuclear fusion—about 10 percent in the center, where temperature and pressure are sufficiently high to fuel hydrogen burning. After about eleven billion years, when the sun has largely depleted the hydrogen in its core, the gas pressure no longer sufficiently supports the residual ashes in the center—which now consist almost entirely of helium atoms—against the gravitational force. Similar to the sun's original contraction from the presolar nebula, gravity again gets the upper hand and squeezes the compact helium core. The core and the layers of unspent gas above it heat up until the burning spherical shell of fresh hydrogen slowly eats its way outward. More and more helium is thus incorporated into the compact ash residue. The temperature and the internal pressure rise substantially, causing the sun's luminosity and diameter to dramatically increase. At the same time, the surface temperature turns slightly cooler, and its color turns reddish. The sun swells into a red giant, the surface of which will ultimately come close to Earth's current orbit.

A little later, the sun will exhale her life as a normal star and expel the remaining hydrogen envelope into a beautiful planetary nebula. These structures are among the most colorful but also bizarre astrophysical objects that the cosmos has to offer (see Plate 24). They form a short, important evolutionary stage at the end of most stars' lives (those with fewer than eight solar masses). Like smoke rings, the red giant star's envelope is ejected at a velocity of tens of kilometers per second into the surrounding medium. Many Hubble images of these glowing gas bubbles reveal very complex structures, such as bipolar rings shaped like an hourglass, several shells possibly dating from earlier eruptions, dust filaments from the atmosphere of the progenitor star, cyclical patterns that look like desert sand dunes, radial spokes that resemble fireworks explosions, or teardrop-shaped cometary structures reminiscent of the proplyds in the Orion Nebula or the Bok globules found in star-forming regions.

The dense ash remaining in the now burned-out fusion furnace in the center of the star will spend the rest of its life as a white dwarf—first as the central star of its planetary nebula and then billions of years into the future after the remaining gas shreds are long gone. This stellar corpse is extremely compact, with atoms so highly compressed they can eventually withstand their own gravity, as Wolfgang Pauli's quantum mechanical exclusion principle demonstrates. Electrons are fermions and, following Pauli, can fill each cell of the phase space only once.[2] A matter form squeezed so that the quantum-mechanically induced pressure of the electrons stands up against the force of gravity is known as a "degenerate electron gas." As a result, a white dwarf star with about half a solar mass is approximately the diameter of Earth. We will encounter white dwarfs again later.

Supernova Explosions

Stars more massive than the sun also emit much more light. A star twice the mass of the sun, for example, radiates about 14 times as bright; a star with ten solar masses is about 6,000 times brighter, and a star with thirty solar masses is approximately 400,000 times brighter. The temperatures on the surfaces of such stars are correspondingly hotter, and their colors appear light blue up to white. A mass-rich star that radiates away its hydrogen supply in such a prodigious fashion obviously must have a much shorter life. A star with two solar masses lives about 1.4 billion years, a star with ten solar masses lives only about thirty-five million years, and all stars of more than thirty solar masses survive only about three million years. Conversely, the life span of red dwarf stars, which are only about one-tenth the mass of the sun, exceeds the age of the universe up to one hundred times.

More massive stars do not end their lives as gently and slowly as our sun but die in gigantic explosions. Like the sun, they initially puff up red giants when the hydrogen in their core is depleted. Once the temperature in the center of the compact ash residue, consisting of helium atoms, increases to about 100 million K, the magic process that Fred Hoyle predicted—and to which we owe our existence—begins. Triple helium atoms fuse to carbon, and as the center heats, nuclear fusion produces the heavier elements. At the center of the star, different elements burn in different layers so that at the end, the core resembles an onion. The phases of this shell burning at the end of a star's life become progressively less efficient and therefore must run faster and faster to allow the star to generate enough energy to withstand the enormous gravitational pressure in its center. For a star of ten solar masses, the fusion of hydrogen into helium takes about ten million years; the burning of helium to carbon and oxygen takes about one million years; the fusion of carbon into heavier elements like neon, sodium, and magnesium takes only ten thousand years; the burning of oxygen to silicon and sulfur takes four years; and finally, the fusion of silicon to iron takes only one week. The nuclear fusion process stops with iron because the formation of even heavier elements does not release energy. On the contrary, additional energy must be supplied. Splitting an iron nucleus also requires energy, making iron the most stable member of the periodic table of chemical elements. This is it for the stellar furnace!

In the last phase, the temperature in the heart of the star rises to about 3 billion K. The dense ball of highly compressed iron steam in the center can no longer withstand its gigantic gravitational force. Even the forces of Pauli's exclusion principle, which stabilize these compact stellar remnants, fail when the mass exceeds a critical limit. Indian-American astrophysicist Subrahmanyan Chandrasekhar calculated this limit in 1930 and received the Nobel Prize in 1983 as a result. The large NASA X-ray observatory Chandra was named after him. Upon reaching the "Chandrasekhar mass," which is about 1.4 times the mass of the sun, the electrons of the degenerate gas zip around inside the compact star almost at the speed of light. Because increasing the electrons' speed cannot compensate for the pressure of the star's own gravity, a compact star with a mass beyond the Chandrasekhar limit inevitably collapses. This happens when the Earth-sized iron ball in the center of the red giant shrinks to a diameter of only twenty to thirty kilometers in a tiny fraction of a second.

During this gravitational collapse, the temperature rises dramatically, to about 100 billion K. The electrons are practically squeezed into the protons of the iron nuclei. This "inverse beta decay" produces a neutron and a neutrino through the weak interaction of one proton and one electron. Within a very short time, the roughly 10^{57} protons and electrons of the iron ball transform into as many neutrons and neutrinos. The collapse stops abruptly when the Pauli exclusion forces of the newly formed degenerate neutron gas halt the gravitational force. A neutron is approximately two thousand times more massive than an electron. In quantum mechanics, the wavelength of a particle, and thus its typical size, gets proportionally smaller with increasing mass. In a degenerate gas, the neutrons, accordingly, require much less space than the electrons. The density in a degenerate neutron gas is approximately the same as that of ordinary nuclear matter. The neutron star formed by this method can therefore be regarded as a single gigantic atomic nucleus with about 10^{57} neutrons and the perimeter of a major city.

In December 1933, just one year after the discovery of the neutron, astrophysicist Fritz Zwicky and his younger, sometimes a little intimidated colleague Walter Baade predicted the existence of neutron stars in two brilliant presentations at the meeting of the American Physical Society at Stanford University, where they introduced the term "supernova." In their original work they wrote:

> With all reserve we advance the view that a supernova represents the transition of an ordinary star into a neutron star, consisting mainly of neutrons. Such a star may possess a very small radius and an extremely high density. . . . A neutron star would . . . represent the most stable configuration of matter as such.[3]

They arrived at this ingenious conclusion by comparing the gravitational energy released during the collapse of a neutron star to the large amount of radiation emitted by a supernova. Robert Oppenheimer, scientific director of the Manhattan Project, and his former postdoctoral researcher George M. Volkoff calculated the first theoretical model of a neutron star in 1939.[4] For many of the physicists involved in the Manhattan Project, supernova explosions were extremely interesting because of their similarity to atomic bomb explosions—and in fact, there are many parallels.

But why does the star explode? Originally, the theoretical astrophysicists, who applied their atomic bomb explosion computer models to the gravitational collapse of a star, believed that the imploding material's impact on the hard surface of the nascent neutron star is sufficient to rip the rest of the star apart. But research eventually showed that the shock wave is just too weak. Since this revelation, attention has focused on the neutrinos, generated in huge quantities during the inverse beta decay, that carry away almost all of the excess gravitational energy. Physics triumphed when supernova 1987A exploded in the Large Magellanic Cloud—after more than three hundred years the first supernova visible to the naked eye—and the large underground detectors, especially the Japanese Kamiokande detector, discovered the theoretically predicted neutrinos. The head of the Kamioka project, Masatoshi Koshiba, received the Nobel Prize for Physics in 2002.

The probability that neutrinos will interact with ordinary matter is incredibly small. One would need a lead shield with a thickness of about one light-year to stop a neutrino. Thus, the neutrinos should be able to leave the dying star practically unimpeded. But just a tiny fraction of neutrino energy would be sufficient to disrupt the whole star. Nevertheless, for decades it remained unclear as to how much energy the neutrinos left in the star during their escape. Although astrophysicists packed more and more detailed neutrino physics into their models, they failed to detonate a star. Even in 2003, my colleague Wolfgang Hillebrandt's research group at the Garching, Germany, Max Planck Institute for Astrophysics titled one of their press releases "Supernova Problem Is Still Unresolved." They were disappointed that one of the most elaborate computer simulations ever performed still failed to produce an explosion.

A few years later, however, they announced a breakthrough.[5] Stars of about ten solar masses now explode in their models. In fact, the neutrinos leave less than 1 percent of their energy in the shock front, which is squeezed between the still infalling matter and the freshly formed neutron star surface. But this energy is sufficient enough to act as a pressure cooker and to accelerate the overlying stellar envelope in a kind of seething supernova explosion (see Plate 19). The blast wave then heats the overlying onion-like shells of heavy elements, melting a portion of the atomic nuclei there into radioactive isotopes of even heavier elements. Supernova 1987A, for example, created large amounts of cobalt and nickel, whose radioactivity could be observed by the X-ray detectors of our institute aboard the Russian space station, Mir.[6]

This could also be the way in which the supernova discovered in 1054 in the Taurus constellation might have exploded.[7] The Chinese emperors of this era employed astronomers to stay constantly informed about important events happening in the sky. Therefore, Chinese historical records about the appearance of new "guest stars" are usually very accurate astronomically, both in terms of the stars' brightness and position. Sources during the Chinese Song Dynasty recorded the emergence of a guest star as bright as Venus on July 4, 1054, and August 27, 1054. The star was supposedly visible for twenty-three days during daylight hours and for almost two years in the night sky. In 1731 and 1758, John Bevis and Charles Messier independently discovered an emission nebula in the Taurus constellation that was later associated with this historic explosion. Messier, after whom the most famous nebulae in the sky are named, discovered this object in his search for Halley's comet and called it Messier 1.

Meanwhile, the remnant of the supernova of 1054 has gained fame under the name of the Crab Nebula, one of the most studied astronomical objects of all time. There certainly must be more than two hundred doctoral theses—including my own—dealing with various aspects of this fascinating object. When I wrote the introduction for my dissertation about thirty years ago, I wondered why only the Chinese had recorded this very bright guest star while not a single European or Arabic reference was known. Had Europe or the Middle East suffered bad weather for several weeks? I was therefore very pleased when, during the research for this book, I learned from the online encyclopedia Wikipedia, which I have begun to cherish, that thirteen historical sightings of this supernova explosion were actually reported: four in Asia (including China), one in Arabia, and eight in Europe. From the historical descriptions—for example, "bright disk in the afternoon," "glowing pillar," "very bright star"—one can even reconstruct the brightness variation and the date of the explosion rather accurately. Hence, the explosion took place on April 11, 1054.

The center of the Crab Nebula contains a fantastic object that crucially influences its appearance. This is a neutron star—the compact remnant of the supernova's progenitor star—that measures approximately thirty kilometers in diameter and has a strong magnetic field about one trillion times larger than Earth's. It rotates about thirty times per second around its own axis. Comparable to a lighthouse spotlight, a highly directional beam of electromagnetic radiation from the radio to the gamma ray range sweeps across the observer twice per revolution, leading to a characteristic periodic fluctuation in the brightness of this "Crab pulsar." The image of the Crab Nebula in visible light (see Plate 20a)[8] on one hand shows the tattered gaseous envelope of the progenitor star in reddish filaments. The diffuse bluish light, on the other hand, represents synchrotron radiation originating from relativistic electrons in the magnetic field of the remnant, which the Crab pulsar has accelerated to the speed of light. Plate 20b shows a photomontage of images zooming in on the Crab pulsar from the Hubble Space Telescope in visible light (red) and from the Chandra Observatory in X-rays (blue).[9] In almost three-dimensional plasticity, we see how the rapidly rotating neutron star ejects fast particles along its equatorial plane and probably also along the polar axis, which are then recognized as annular and jet-like structures. My colleagues Bernd Aschenbach and Wolfgang Brinkmann at the Max Planck Institute for Extraterrestrial Physics (MPE) in Garching, Germany, theoretically predicted this geometry of the acceleration region in 1975[10]—long before it could actually be observed, first with the Roentgen Satellite (ROSAT) and later with Chandra.

Explosions as Standard Candles

In addition to the supernova explosion that occurs when a massive star's iron core collapses, another important type of stellar explosion, to a certain extent, originates on the surface of a white dwarf star. These are the "type Ia" supernovae, which have played a particularly important role in cosmology and in the discovery of dark energy. Many stars are born in binary systems. Because they usually have different masses, they also evolve on different time scales. The more massive of the two stars begins first to swell into a red giant and later turns into a white dwarf. Sometime later, after the companion star develops into a red giant, and the two stars are close enough to each other, part of the expanding envelope of the red giant can fall onto the white dwarf in a process called "accretion." As matter falls down to the compact dwarf, it heats up dramatically and starts to radiate ultraviolet light and X-rays. Often, these accretion processes vary to the extreme over time. Dramatic bursts of X-ray and ultraviolet radiation alternate with quiet periods. Accreting white dwarfs in binary systems are therefore also referred to as "cataclysmic variables."

The fresh hydrogen rained down upon the surface of the white dwarf wraps around the compact star as a new shell. When the pressure, density, and temperature in this shell grow sufficiently large, the hydrogen may begin to burn and fuse into helium. Often, this happens explosively, similar to the ignition of a hydrogen bomb. In this case, the shell is blown away from the star and becomes brighter as it inflates. For a short time, therefore, a new star, a nova, appears in the sky. Compared to a supernova, in which the entire star explodes, a nova involves only the thin shell and thus creates much less energy.

Using ROSAT, in 1991 some of my colleagues and I were able to identify a new class of ultraluminous X-ray sources with very soft X-ray spectra—so-called "supersoft" sources (SSS).[11] Theoretical studies show that in these sources, the hydrogen burns continuously on the surface of white dwarfs. We therefore can directly observe the nuclear fusion that otherwise remains hidden in the interiors of the stars.

Because in the case of these SSS the accreted matter is not thrown back into space but accumulates on the surface of the white dwarf, the star becomes more and more massive. Interestingly, the size of the star actually gets smaller, instead of larger, during this process due to the quantum mechanical properties of the degenerate electron gas. This continues until the star reaches the critical Chandrasekhar mass of 1.4 solar masses, at which point the electrons become relativistic and cannot withstand the massive gravitational force. However, in contrast to the iron core collapse in the massive stars discussed previously, the white dwarf still contains many lighter elements such as carbon, which can generate fusion power. As a result, a thermonuclear runaway explosion occurs, similar to a hydrogen bomb. But in this case the entire star explodes, and all of its components shoot into space.

Because fundamental properties of matter determine the Chandrasekhar mass, these type Ia explosions are almost identical from star to star and always produce approximately the same amount of energy.[12] Therefore, such explosions can be used as "standard candles" that allow us to explore the geometry of the universe. In 1998, measurements of distant supernova explosions provided important clues to the existence of dark energy, which accelerates the expansion of the cosmos (see Chapter 2). Two teams who independently discovered the cosmic acceleration received the Nobel Prize for Physics in 2011. University of Hawai'i astronomer John Tonry, who introduced digital wide-field Charge Coupled Device (CCD) measurement techniques used on Maunakea into the supernova research, was a member of one of those teams and was invited to the Stockholm Nobel celebrations. In 2015, he also shared the prestigious "Breakthrough Prize" with his colleagues.

THE BIGGEST EXPLOSIONS AFTER THE BIG BANG

In the 1960s, at the height of the Cold War, the superpowers of the United States and the Soviet Union agreed to ban nuclear testing in the atmosphere. In order to monitor this test moratorium, and also to ensure that no nuclear tests were performed in space—not even on the backside of the moon—the United States launched a number of Vela series satellites. With an orbit of up to 250,000 kilometers, these satellites orbited in pairs on opposite sides of Earth, so they always kept the whole world in view and could monitor compliance with the moratorium using X-ray, gamma ray, and neutron detectors. In fact, they discovered a number of events associated with nuclear testing, which resulted in political tensions. Most importantly from an astronomical point of view, however, was the discovery of several mysterious gamma ray bursts that seemed to originate not from Earth, nor from the sun or the moon. In 1973, Ray Klebesadel from the Los Alamos Laboratory announced the discovery of cosmic gamma ray bursts, based on the Vela satellites. At the same time, colleagues on the other side of the Iron Curtain, led by Evgeny Mazets from the Ioffe Institute in Leningrad, also observed gamma ray bursts with the Russian satellite Cosmos 461. According to these measurements, approximately once a day somewhere in the sky, a flash of gamma rays occurs for seconds to minutes that is brighter than the entire gamma sky.

In subsequent years, astronomers quarreled vividly about the nature of gamma ray bursts. Most believed that we were seeing objects inside our Milky Way, for example, neutron stars, which release explosive energy during starquakes.[13] A minority suspected that these gigantic explosions occurred in other galaxies very far away, while other astronomers wanted to place the explosions within our own solar system—for example, in the Kuiper Belt. If we study the distribution of gamma ray bursts in the sky and their brightness, different scenarios can, in principle, be ascertained. If they originate in the Milky Way, for example, the gamma ray bursts should be distributed anisotropically over the sky, with an accumulation in the direction of the Galactic Center since the sun is located in the outer regions of the Milky Way. The X-ray instrument Burst and Transient Source Experiment (BATSE) on the NASA Compton Gamma-Ray Observatory (CGRO) mapped a sufficient number of gamma ray bursts in the early 1990s to clearly prove that their distribution is perfectly isotropic over the sky, thus virtually eliminating a galactic origin. Throughout the years, however, these gamma ray bursts could only be localized very crudely, and it was impossible to associate them with specific celestial objects visible in other wavelengths. The breakthrough came with the measurements of the Italian-Dutch satellite BeppoSAX, which for the first time could detect the faint X-ray afterglow of a gamma ray burst. The BeppoSAX team utilized a clever observation strategy that quickly analyzed the signals of the wide-angle X-ray camera in the ground station and then immediately commanded a slew of the satellite to observe in the proper direction using its narrow-field X-ray telescopes. On February 28, 1997, the team determined the location of a burst in the sky so accurately that ground-based optical telescopes were able to observe the afterglow.[14] Approximately twenty-one hours after the onset of the burst, Dutch astronomer Jan van Paradijs and his colleagues discovered a rapidly fading optical object near a distant galaxy, thus proving that gamma ray bursts have a cosmological origin.[15]

On November 20, 2004, NASA launched the Swift satellite, which is dedicated to hunting gamma ray bursts. Routinely and autonomously, within seconds the satellite performs on-board analyses and operations that previous missions had to run at the ground station. Thus, Swift has revolutionized the analysis of gamma ray bursts. A whole swarm of astronomers check on the pulse of Swift around the clock through automatic alarm messages on their cell phones. As of April 2015, there were a total of 1,459 bursts. For 966 of those bursts an afterglow was measured in X-rays, 607 were discovered in visible light, and 109 were seen in radio waves.[16] For a total of 427 bursts, redshifts, and therefore distances, have been determined. The galaxies that host gamma ray bursts are among the most distant objects in the universe. The two record-holding gamma ray bursts, at redshifts of 9.4 and 8.2, were observed during one week in April 2009. Their light was sent out when the universe was 500 to 600 million years old and about one-tenth its current size.

Despite these huge distances, gamma ray bursts appear very bright in the sky and must process huge amounts of energy. In a time interval of fractions of a second up to a few minutes, these bursts radiate the total energy content of a star into space. They are therefore also referred to as the "biggest explosions after the Big Bang." Such explosion energies can only be liberated during the formation of a stellar mass black hole. Theoretical analysis—the so-called fireball model—indicates that relativistic particle beams play a crucial role in gamma ray bursts. Layers of matter are ejected from a dying star and collide with each other at almost the speed of light, generating strong gamma radiation. A connection with hypernovae—explosions that are even more gigantic than supernovae—has been discovered for a number of gamma ray bursts. Therefore, we assume that we are witnessing the birth of black holes from the core collapse of very massive stars. In another subclass of gamma ray bursts that only lasts for fractions of a second, we may see a stellar black hole born out of the collision of two neutron stars. Justifiably, we can therefore argue that this is the birth of the "cosmic monster" that we will study in more detail in Chapter 8.

Chapter Eight

COSMIC MONSTERS

Toward the end of the eighteenth century, two researchers considered the effect of gravity on light. According to Newton's gravity laws, every object has a so-called escape velocity that must be exceeded in order to escape its gravity field. If you throw a stone up into the air, it will reach a certain height and then fall back to the ground. The more energy—and thus initial velocity—that you put behind your throw, the higher the stone will rise. No human power could throw a stone away from Earth, which has an escape velocity of about 4,000 kilometers per hour, or 11.2 kilometers per second. Only a rocket propulsion system can achieve such high speeds—so if you buckle a sufficiently large rocket under a stone, it can leave Earth.

The sun's escape velocity is 617 kilometers per second. The escape velocity is dependent on both the mass and the size of the object. The denser and smaller an object is, the higher its escape velocity.[1] In 1784 and 1796, English naturalist Rev. John Mitchell and French mathematician and astronomer Pierre-Simon Marquis de Laplace independently theorized the properties that a star whose escape velocity corresponded exactly to the speed of light would possess. Not even light could escape from such a star; it would be completely dark. The speed of light is almost exactly 300,000 kilometers per second.[2] One of Mitchell's or Laplace's "dark stars" with the mass of the sun would therefore have to be 236,000 times more compact than the sun. Because the sun has a radius of 696,000 kilometers, this turns out to be

143

three kilometers. Imagine the entire mass of the sun squeezed into the diameter of an average small town! On the other hand, the dimensions of neutron stars with radii of about 10 to 15 kilometers and solar masses of about 1.4 are not much different. Neutron stars have escape velocities that are a sizable fraction of the speed of light.

Mitchell's and Laplace's predictions were way ahead of their time. Even after more than one hundred years, their forgotten considerations should be rediscovered. The brilliant astrophysicist Karl Schwarzschild played a crucial role in this story. Schwarzschild was born in Frankfurt am Main in 1873 and became thrilled by astronomy and physics at a young age. As a sixteen-year-old student, he published two papers in the renowned scientific journal *Astronomische Nachrichten*. By the age of twenty-seven, he was professor and director of the Göttingen Observatory. By that time, he had already published groundbreaking work on determining the intensities on photographic plates. In Göttingen, he worked with great mathematicians such as David Hilbert and Hermann Minkowski. There, he also hired Danish astronomer Ejnar Hertzsprung, the coinventor of the Hertzsprung-Russell diagram, which plays a very important role in understanding stars. In addition to many other seminal astrophysical studies, Schwarzschild developed the theory of geometrical optics, which forms the basis for all of today's mirror telescopes.

In 1909, Schwarzschild was appointed director of the Astrophysical Observatory Potsdam, where he reluctantly moved, taking the recently hired Hertzsprung along. In 1912, he was appointed to be a member of the Prussian Academy of Sciences. Max Planck held the inaugural speech for his admission. In the same year in Potsdam, his son Martin Schwarzschild was born. Martin also became a famous astrophysicist, and I had the pleasure of meeting him in person at Princeton University about eighty years later.

At the outbreak of the First World War in 1914, Schwarzschild, feeling obligated to his country as a German Jew, immediately volunteered to join the army. Although he went to the front, he still found the time to work on physical questions. He sent most of his work from this time period directly to Albert Einstein in Berlin, who presented it in Schwarzschild's name at the Academy of Sciences. On the Russian front in December 1915, just weeks after Einstein's publication of the theory of general relativity, Karl Schwarzschild found the first exact solution to Einstein's field equations. He calculated the curved space-time outside an electrically neutral, nonrotating, spherically symmetric star in the vacuum. His preliminary "outer" solution, "on the gravitational field of a mass point according to Einstein's theory," later given the name "Schwarzschild geometry," was sent to Einstein. Einstein was surprised at the speed with which the solution had been found and congratulated Schwarzschild on its simplicity. In January 1916, Einstein presented the results on Schwarzschild's behalf at a meeting of the Prussian Academy of Sciences. According to this theory, there is a critical distance for any mass, within which the curvature of space becomes so large that not even light can escape. Interestingly, the "Schwarzschild radius," named in his honor, corresponds exactly to the size Mitchell and Laplace discovered more than one hundred years earlier. For the mass of the sun this radius, as mentioned above, is about three kilometers; for Earth it would be one centimeter.

Only a short time later, Schwarzschild submitted his "inner" solution to Einstein's equations for a star filled homogeneously with matter. A few weeks later, he was infested at the front by a heavy skin disease that was incurable at that time. He returned to Potsdam and died in May 1916 at the age of only forty-two. Einstein gave the eulogy at his funeral.

Although Schwarzschild had calculated his first solution only for the exterior of a star, the Schwarzschild geometry is now generally applied to a point mass in a vacuum. Schwarzschild's "trick" was to tinker with the coordinates of space and time—that is, to apply correction factors to both variables, which grow in importance as you get closer to the Schwarzschild radius. Kip Thorne, the famous explorer of black holes at the California Institute of Technology, therefore titled his vividly written book *Black Holes and Time Warps*.[3] Inside the Schwarzschild radius, the coordinates of space and time swap properties in a bizarre way: time becomes "space-like" and space becomes "time-like," which for every coordinate inside the star inexorably leads to a so-called singularity—a mass point at which the density and the curvature of space become infinite.

Throughout his entire life, Einstein refused the construct of the "Schwarzschild singularity" as illogical and unphysical. He believed there must be a law of nature that prohibits such nonsense and escaped, as it were, the predictions of his own theory. Even today, most physicists do not believe that singularities exist. If one were to venture to smaller and smaller dimensions, one would inevitably at some point arrive in the regime of quantum physics. If, in addition, large gravitational forces are at play, the physics of quantum gravity becomes important, which nowadays no one really understands. In the string theory, one of the most popular candidates for a quantum theory of gravity, the singularity within the Schwarzschild radius is replaced by a kind of "spaghetti ball" of strings tangled with one another.

Even after Schwarzschild's solution, it still took several decades of scientific debate among great physicists such as Eddington, Chandrasekhar, Oppenheimer, and Landau until scientists accepted the reality of objects so compact that within their vicinities, relativistic effects dominate, and whose escape velocity is close to the speed of light. As we saw in Chapter 7, Walter Baade and Fritz Zwicky predicted the existence of neutron stars before the Second World War. These stars should possess about 1.5 to 2 times the mass of the sun but are only ten to twenty kilometers in radius and are stabilized by quantum mechanical forces. In the 1960s, radio and rocket technologies invented during World War II and finally adapted to scientific purposes made possible three important discoveries establishing the field of "relativistic astrophysics": in 1962, Riccardo Giacconi and his colleagues discovered the compact X-ray star Scorpius X-1, the brightest X-ray source in the sky, which turned out to be a neutron star accreting matter from a companion star, as well as the diffuse cosmic X-ray background. In 1963, Maarten Schmidt identified the quasars—very energetic compact objects that can outshine the center of their galaxies thousands of times. In 1968, Jocelyn Bell discovered the radio pulsars—rapidly rotating neutron stars whose light sweeps over the observer like a lighthouse beam. In the same year, US astrophysicist John Archibald Wheeler coined the term "black holes" for Mitchell's and Laplace's dark stars. Shortly thereafter, measurements from the NASA X-ray satellite Uhuru clearly indicated the existence of such compact objects. In 1970, the mass of the compact object in the X-ray binary system Cygnus X-1 could finally be determined. It turned out to be the best candidate for a stellar-mass black hole.

Until a few years ago, it was assumed that the properties of black holes depend mainly on their mass and only two other physical quantities, namely, their electric charge and their angular momentum. Hans Reissner and Gunnar Nordström calculated the solution to the Einstein equations for an electrically charged black hole independently of each other in 1916 and 1918. In practice, however, one assumes that the electric charge of black holes should approximate zero because it is very difficult in nature to separate large electrical charges. Much more important for practical applications is the geometric solution for rotating black holes that New Zealand mathematician Roy Kerr discovered in 1963. The rotation of black holes plays an important role in astrophysics and is associated with the emergence and acceleration of highly collimated jets of particles. In addition to the three quantities of mass, charge, and angular momentum, a black hole was thought to have no other physical properties. This situation was described by the theorem: "Black holes have no hair."

But this is only true if we limit the consideration to the theory of relativity. Taking quantum mechanical effects into account yields interesting surprises. As we have already noted, the two great theories of the twentieth century do not fit together. Physicist Stephen Hawking spent decades thinking about how to bring the two theories together. The quantum mechanical effects he discovered in the vicinity of black holes could be a first clue to a theory of quantum gravity. Hawking considered the vacuum in which the black hole resides. As we saw in Chapter 1, the vacuum should always be filled with quantum fluctuations, those virtual particle pairs that occur for a short time out of nowhere and immediately disappear again. If any of these virtual pairs is produced near a black hole, the black hole could swallow one of the two particles while the other, suddenly without its partner, moves away. A virtual particle can thus be transformed into a real particle. The newly created particle extracts a small amount of energy from the environment of the black hole, which in turn needs to balance the energy loss by minimally reducing its radius and thus its mass. We imagine the emergence of real particles from quantum fluctuations occurred in a similar way during the inflation phase at the beginning of the universe.

Hawking proposed the revolutionary theory that a black hole radiates away energy. However, this is a tiny amount of energy. Hawking calculated that such loss processes most likely occur when the virtual particle pair has about the wavelength (or size) of the Schwarzschild radius. In a much smaller or larger particle pair, it is unlikely that a single partner falls into the hole because either both or neither would be swallowed. The mean wavelength of the Hawking radiation therefore corresponds precisely to the Schwarzschild radius. This way the radiation power that a black hole emits, as well as its lifetime, can be calculated. The bigger a black hole is, the longer it lives. A black hole the mass of the sun lives for about 10^{67} years. Conversely, small black holes only live for a very short period of time. At the end of their lives, they diminish and quickly evaporate in a gamma ray flash. From this, we can conclude that no such small black holes exist in the universe today, otherwise we would see their gamma ray emissions.

Hawking radiation is also interesting in another context. Before the Large Hadron Collider (LHC) large accelerator at the European Organization for Nuclear Research (CERN) started operation, it was speculated that due to the huge energy density of its particle collisions, tiny black holes could spontaneously form in the interaction zone. Therefore, environmentalists protested against this new generation of accelerators, fearing that the mini-black holes would grow very quickly by swallowing matter and in a relatively short time, drill their way to the center of Earth. They might possibly even eat Earth from the inside out! However, it can be shown theoretically that the Hawking radiation of a small black hole is so intense that it would almost immediately evaporate before causing any serious calamity. Even more reassuring is the fact that cosmic rays at the highest energies can trigger the same reaction in Earth's atmosphere, and obviously, no dangerous black holes have appeared in 4.6 billion years. When the LHC went live in September 2008, fortunately, no black holes materialized, and the whole storm in a teacup calmed down. However, those same environmentalists still argue that there might be a danger during the LHC's higher energy phase that started in 2015.

Hawking, however, went one step further. He wondered what happens to the information that falls into a black hole. If you throw a book into the fireplace and then follow the exact movements of all particles involved through the fire, the ash, the smoke, and the radiation, you would in principle be able to reconstruct the contents of the book. The information is conserved in nature. Not so with Hawking's black holes. According to the "no hair" theorem, the Hawking radiation should be completely unstructured. If you wait long enough, the mass accumulated by the black hole would be radiated back into the universe—but not the swallowed information. This led Hawking and Kip Thorne to put forward the idea of "white wormholes." The center of a black hole could be connected via an unfathomable kind of space-time loop with another universe or with another area of our own universe, where it would spit out the gathered information as a white wormhole. This idea inspired quite a number of science-fiction authors to write stories about time travel through black holes toward other universes. Kip Thorne even speculates in *Black Holes and Time Warps* that one might be able to create artificial black holes and white wormholes in order to move from place to place at incredible speed.

In 1997, Stephen Hawking and Kip Thorne made a public bet against their colleague John Preskill. Hawking and Thorne felt that the information would disappear irretrievably, while quantum theorist Preskill remained convinced that a correct quantum gravity theory would prove that the information would ultimately come back out of the black hole. The winner would get an encyclopedia of his choice. In summer 2006, at a meeting of the leading relativity and gravitation researchers in Dublin, Stephen Hawking announced with great fanfare that he had been wrong with his original analysis and that black holes were not "skinheads." The information would be delivered back again with the Hawking radiation via microscopic structures on the Schwarzschild radius. He apologized to all sci-fi fanatics that therefore, there are no passages into other universes and, as a result, no time travel. Kip Thorne, however, was still skeptical and refused to honor the joint bet. In the same year, Samir Mathur, a physicist at Ohio State University, used string theory to determine that the interiors of black holes do consist of the above-mentioned spaghetti ball of strings. These punch through the Schwarzschild radius and therefore, black holes have "hair" indeed.[4]

THE MONSTER IN THE CENTER OF THE MILKY WAY

As can be seen in Figure 8.1, we know of two types of black holes: stellar-mass black holes of about ten solar masses and supermassive black holes. The latter, gigantic objects ranging from millions to billions of solar masses, are located in the centers of galaxies. The most exciting and best-studied supermassive black hole is located in the center of our own Milky Way, a complex and fascinating area. In a small crowded space, we find clouds of infalling gas and dust, hot gas from earlier supernova explosions, and a star cluster with several thousand young stars. But the most interesting object is a black hole that holds about 4.4 million solar masses. It therefore has a Schwarzschild radius of about thirteen million kilometers, only about nineteen times the solar radius. Imagine an object that is four million times the mass of the sun but only nineteen times larger! The average density in this eerie gravity trap is about nine hundred times higher than that of the sun.

The center of the Milky Way is hidden from our view behind thick, dark clouds of gas and dust (see Figure 1.1). Only radio waves, infrared light, and X-rays can penetrate these walls. In the mid-1970s, radio astronomers discovered the first indications of a bizarre object in the Galactic Center: a bright, compact radio source in the constellation

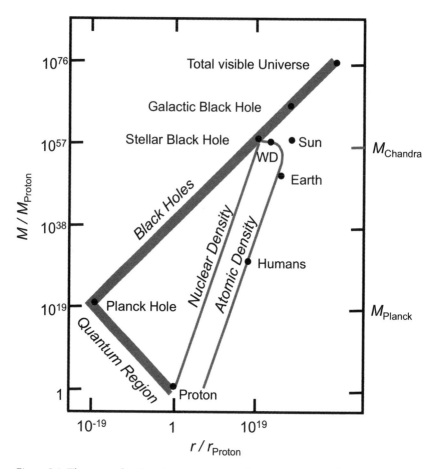

Figure 8.1. The mass of various known structures in space (in units of the proton mass) plotted against their size (in units of the proton radius). Objects, which in essence are held together by electromagnetic forces, together with people and planets, lie on a single line given by the density of normal atoms. Objects with the density of atomic nuclei, in particular the proton itself, lie on the line labeled "Nuclear Density." Stellar black holes as well as neutron stars remain close to the density of nuclear matter. The Schwarzschild radius, beyond which an object turns into a black hole, is denoted correspondingly with a thick diagonal line. It is proportional to the mass of an object. The "Quantum Region" refers to the Compton wavelength and thus the characteristic size in which the quantum nature of an object or a particle dominates. It is inversely proportional to the mass of the particle. The point where the Schwarzschild radius and the Compton wavelength meet is the Planck region described in Chapter 1. The "Planck hole" thus represents the smallest possible black hole. (After Carr and Rees, 1979, *Nature* 278, p. 605.)

Sagittarius, which they called Sagittarius A* (Sgr A*). Back then, astronomers conjectured that it probably represented the exact center of our galaxy and was very likely a black hole. This radio source is embedded in a region with fascinating gas structures. A minispiral and sharp filamentary radio arcs suggest that gas masses plunge into the center at great speeds, where strong magnetic fields squeeze them together.

Just like X-ray and radio emission, infrared radiation is able to penetrate dense gas and dust clouds. Large telescopes can therefore take exciting pictures of the center of the Milky Way in these spectral ranges. However, combatting turbulence in the atmosphere presents a major challenge, which means that the images of the stars twinkling in the sky dance around on the telescope images and therefore appear blurred. Astronomers call stars' flickering in the sky "seeing." It can be compared to the summery shimmer over a hot asphalt road. Due to the rapid technological development in optical technologies, it has become possible to take sharper and sharper infrared images in the last two decades; first, with "speckle methods" that trick the air turbulence and correct it with high-speed recordings, and second, with "adaptive optics" that compensate for the scintillation of the atmosphere using a fast-moving "rubber" mirror and ultimately even an artificial laser-guided star. Ideally, the images are now only limited by the diffraction of the telescope aperture, enabling the large ground-based telescopes with 8- to 10-meter mirrors to achieve clearer images than the 2.2-meter Hubble Space Telescope.

A group headed by my colleague Reinhard Genzel at the Max Planck Institute for Extraterrestrial Physics (MPE) in Garching, Germany, leads these technological developments, and since 1992 has observed the Galactic Center using increasingly sophisticated methods and telescopes at the European Southern Observatory (ESO) in Chile.[5] A little later, a group led by Andrea Ghez of the University of California, Los Angeles joined the competition with measurements using the Keck telescopes in Hawai'i.

With the help of diffraction-limited infrared images, astronomers have successfully observed a cluster of luminous stars in the immediate vicinity of the Galactic Center (see Plate 21a). Interestingly, these stars are surprisingly young. They must have formed together about five million years ago, just at the time on Earth when the first hominids broke away from the other primates. It is still a mystery how these stars could have developed in the inhospitable environment of the

Galactic Center. This is an active research topic pursued, among others, by our Institute for Astronomy faculty member Jessica Lu.

Years of detective work have proven that these stars actually swirl around the object at the position of Sgr A*, barely visible in the infrared, with amazingly high speeds, some at more than one thousand kilometers per second. The closer the stars are to the center, the faster they move. Astronomers have practically repeated at the Galactic Center what Johannes Kepler and his contemporaries showed us four hundred years ago with the planets in the solar system. According to Kepler's laws, a body (such as a planet or a comet) moves in the gravitational field of its star in an elliptical orbit. One of the two focal points of the ellipse is the system's center of mass. On the segment of its orbit furthest away from the center of gravity, the object moves slowest. It reaches the highest speed at the point of its path closest to the gravity focus (the pericenter). The only difference in the Galactic Center is that instead of a single star controlling the orbit of its planets, it is a monster of almost four million solar masses flinging the stars in its vicinity into elliptical orbits. In spring 2002, the Genzel group made a dramatic discovery. One of the stars closest to the center—bearing the prosaic name Star 2 (S2)—could be observed in the immediate vicinity of its closest flyby. Its speed was about ten thousand kilometers per second, roughly 3 percent of the speed of light! Both groups in the United States and in Europe continue to observe S2, which has, in the meantime, performed a complete orbit around the Galactic Center. This is the first Kepler orbit observed in the sky outside our own solar system.

So far, more than two dozen stellar orbits have been tracked down in the Galactic Center (see Plate 21b).[6] Using integral-field spectroscopy that can simultaneously measure spectra for all the objects in an image, astronomers have even succeeded in capturing three-dimensional information. As a side effect of these observations, they can determine the distance to the Galactic Center with unprecedented accuracy. The fact that all these orbits run exactly around the same center allows astronomers to make an important inference about the maximum size of the central compact object. In principle, it would be possible to hide the four million solar masses not in a single compact object, a black hole, but in a very compact cluster of millions of neutron stars or stellar-mass black holes. However, the measurements by Reinhard Genzel and Andrea Ghez, as well as the phenomena described below, are putting these thought experiments to rest. The massive object in the Galactic Center is just too small to be a star cluster. The only possible alternatives to a black hole include a boson star or a Grava-star made of dark energy, which appear too far-fetched and exotic to most astrophysicists. Reinhard Genzel, who approaches these matters very cautiously and over the years has consistently avoided the term "black hole," has become one of the most powerful advocates for black holes. However, we must keep in mind that all astronomical measurements only refer to the area outside the Schwarzschild radius; that is, to the effect that black holes exert on matter and radiation in their vicinity. Nothing will allow us to look inside the Schwarzschild radius. So when theorists make statements about the interior of a black hole—for example, to distinguish between a singularity, string spaghetti, or quantum foam, that's fine, as long as the outer space of these objects behaves as Karl Schwarzschild predicted.

FEEDING THE MONSTER

Of course, the Galactic Center has long been the focus of X-ray astronomers. An observation with the NASA Einstein Observatory in the 1980s showed that only very weak X-rays, if any, emanate from the center. About ten years later, the German X-ray Roentgen Satellite (ROSAT) discovered a tentative, weak source at the Sgr A* site. But only the current NASA and European Space Agency (ESA) X-ray satellites Chandra and XMM-Newton have been able to prove the existence of X-ray radiation from the immediate vicinity of the black hole beyond any doubt.

Why does a black hole radiate? Should it not swallow everything, including the light? In reality, we do not see the black hole itself; merely the matter in its immediate vicinity, which is swallowed by this maelstrom. This matter can reach velocities near the speed of light and heats up due to friction losses in the accretion disk (Plate 22) as well as through other high-energy processes that have yet to be fully understood. It therefore reaches such high temperatures that it starts to radiate strongly, particularly in X-ray light. We see this as if it were the matter's "last cry for help" just before it falls into the black hole. In addition to light, the black hole's environment often ejects matter particles. Sharply focused particle beams, called jets, are accelerated to relativistic speeds. The jets emit light in a broad range of the electromagnetic spectrum, from radio to gamma rays (Plate 23).

In 2004, Chandra and XMM-Newton discovered X-ray outbursts from the black hole, during which Sgr A* got dramatically brighter— typically for less than half an hour at a time. Over a period of about fifteen minutes, the X-ray radiation increased up to one hundred times, and the drop in intensity occurred just as quickly. Only a single compact object of the corresponding mass—a black hole—can be switched on and off again so rapidly. Sometime later, similar flares were found in the infrared range. Interestingly, both the X-ray and the infrared flares varied significantly over a time scale of about ten to twenty minutes, allowing inferences about the size of the black hole. These processes must take place in the immediate vicinity of the Schwarzschild radius. After monitoring the black hole with sensitive telescopes, scientists realized that such outbreaks happen about once a day. One can say that every day the black hole takes a little "snack." How much mass such a meal contains can be estimated from the X-ray and the infrared luminosity of such an eruption: every time, an object roughly the size of a comet or a mountain is swallowed!

On the other hand, the black hole is surprisingly quiet in the time between outbreaks—so quiet that it proves to be quite a headache to astronomers. Let us just imagine the massive, young stars orbiting the black hole: winds of charged gas and dust particles must blow from the surface of these stars at forces much stronger than the solar wind blowing toward our Earth. At some point, the black hole should capture the material in these winds, and they should emit light, as described above. But in its quiet phase, the black hole emits one hundred thousand times less than what would be expected. We still do not fully understand this extremely inefficient accretion process. Probably, the angular momentum mentioned previously plays a role, leading the material orbiting around the black hole to accumulate in a disk. Like the rings of Saturn, the matter does not even think to fall down to the central mass because it stays balanced between gravity and the centrifugal forces. This remains true as long as the material in the disk is spread so thin that the particles hardly interact with each other. If, on the other hand, the density in this disk grew high enough, the matter would form a dense molasses due to mutual friction and probably local magnetic fields. It would heat up due to friction losses and slowly move inward. But apparently, this is not the case in the Galactic Center.

In 2012, the discovery of a gas cloud slowly moving toward the Galactic Center spurred a lot of excitement. More food seemed to be arriving for the black hole! Its pericenter passage was predicted for a

PLATE 14. Star formation in the Rosetta Cloud, as observed with the European Space Agency's (ESA) Herschel Observatory. (Courtesy of ESA/PACS and SPIRE Consortium/ HOBYS Key Programme Consortia.)

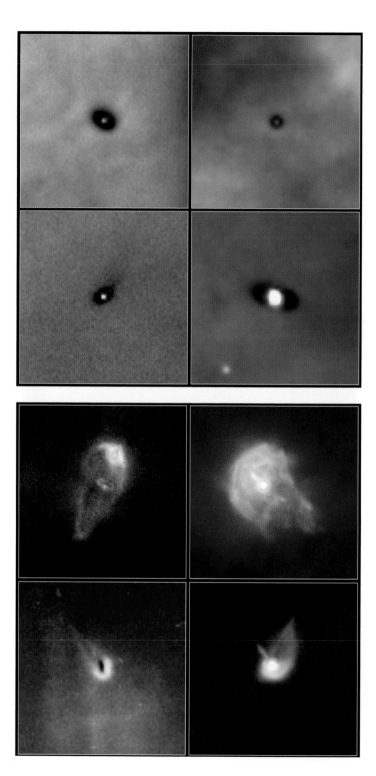

PLATE 15A; 15B. *Opposite page:* Hubble Space Telescope images of protoplanetary disks, at a distance of about 1,500 light-years, in the star-forming region of the Orion Nebula. *Top:* These four images show gas and dust disks, which appear as shadows in front of the bright background of the Orion Nebula. (Courtesy of Mark McCaughrean, Max Planck Institute for Astronomy; C. Robert O'Dell, Rice University; and NASA.) The young stars in the center of these disks, which have already freed themselves from the surrounding cocoon, are directly visible. *Bottom:* These four pictures show protoplanetary disks—"proplyds" in the abbreviated form, which are being harassed by the bright stars in the Orion Nebula. (Courtesy of NASA; European Space Agency; J. Bally, University of Colorado; H. Throop, Southwest Research Institute; and C. Robert O'Dell, Vanderbilt University.) The comet-like or tadpole-like shape comes from the fact that the bright stars in the Orion Nebula blow away the material of the proplyds with their strong ultraviolet radiation and stellar winds.

PLATE 16. *Below:* Young stars are forming at the top of dust pillars in the Carina Nebula. (Courtesy of NASA, the European Space Agency, M. Livio, and the Hubble Twentieth Anniversary Team/Space Telescope Science Institute.)

PLATE 17. Artist's conception of the protoplanetary disk around the star LkCa 15. The "baby planet" LkCa 15 b is still accreting material from its cradle of gas and dust. (Courtesy of K. Teramura, University of Hawai'i.)

PLATE 18. Artist's conception of the "lava planet" Kepler 78b. (Courtesy of K. Teramura, University of Hawai'i.)

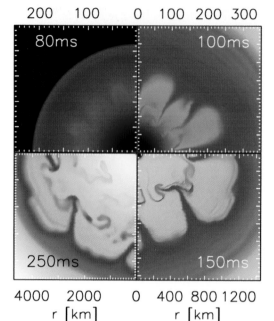

200   100   0   100 200 300

80ms   100ms

250ms   150ms

4000   2000   0   400 800 1200
    r [km]   r [km]

PLATE 19. Computer simulation of the explosion of a star of about ten solar masses. Instabilities between the neutrinos pushing outward and the gravity pressing down on the core are blown away by the explosion and are probably responsible for the fascinating filamentary structures of supernova remnants. (Courtesy of Max Planck Institute for Astrophysics.)

PLATE 20A; 20B. *Opposite page:* The Crab Nebula *(top)* observed with the FORS2 instrument of the Very Large Telescope. (Courtesy of the European Southern Observatory.) This nebula is the remnant of a supernova that exploded in 1054 about six thousand light-years away in the constellation Taurus. The bluish emission is synchrotron radiation from the high-energy electrons accelerated in the center of the nebula. The reddish filaments originate from the compressed gas envelope of the progenitor star. *Bottom:* In the center of the nebula is a neutron star, a so-called pulsar, rotating about thirty times per second around its axis and thereby emitting electromagnetic radiation from radio to gamma rays like a lighthouse. The image is composed of a picture from the Hubble Space Telescope in visible light *(red)* and an X-ray image from the Chandra Observatory *(blue)*. (Images: X-ray: NASA/CXC/ASU/J. Hester et al.; Optical: NASA/HST/ASU/J. Hester et al.)

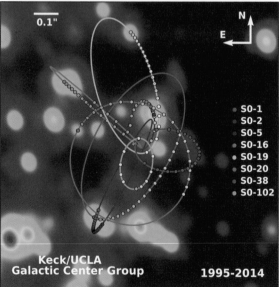

PLATE 21A; 21B. *Top:* An infrared image of the Galactic Center obtained by the group of Reinhard Genzel using adaptive optics with the Very Large Telescope of the European Southern Observatory. (Courtesy of the Max Planck Institute for Extraterrestrial Physics.) *Bottom:* Illustration of the paths that describe the orbits of the innermost stars around the black hole Sgr A*. (Courtesy of A. Ghez, University of California at Los Angeles.)

Within the bottom image:

0.1"

N
E

S0-1
S0-2
S0-5
S0-16
S0-19
S0-20
S0-38
S0-102

Keck/UCLA
Galactic Center Group

1995-2014

PLATE 22. Artist's impression of a black hole. (Courtesy of the European Space Agency; J. Wilms, NASA.)

PLATE 23. The gargantuan galaxy M87 in the center of the Virgo Cluster contains a black hole of about three billion solar masses and ejects a jet of matter far into the universe at almost the speed of light. (Courtesy of NASA and the Hubble Heritage Team, Space Telescope Science Institute/Association for Universities for Research in Astronomy.)

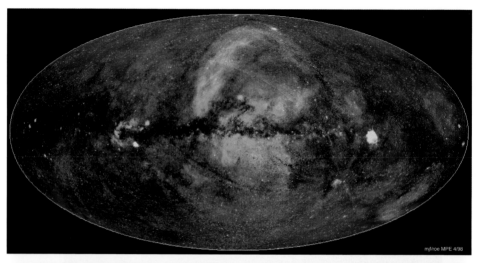

mjf/roe MPE 4/98

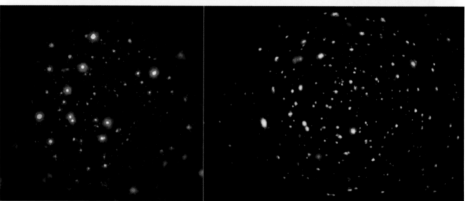

PLATE 24. *Top:* Panorama of the entire sky in X-rays as measured by the Roentgen Satellite (ROSAT). The X-ray radiation is color-coded. *Red:* Soft radiation with temperatures of about one million K. *Green or blue:* Correspondingly harder radiation with temperatures in the range from three to ten million K. In the middle of the figure is the Galactic Center. (Courtesy of the Max Planck Institute for Extraterrestrial Physics.)

PLATE 25. *Botttom:* Deep X-ray survey in the Lockman Hole. The field is about as large in the sky as the full moon. *Left:* The longest (1.4 milliseconds) and most sensitive observation with the Roentgen Satellite (ROSAT) in false color. The data recorded with the color-sensitive PSPC are *red* (very "soft" X-rays, 0.1–0.5 keV) and *blue* ("soft" X-rays, 0.5–2 keV). *Green:* Data from the high-resolution HRI camera. *Right:* The X-ray color image of the same sky region recorded with XMM-Newton using an exposure time of about one millisecond. The colors correspond to the energy ranges. *Red:* "Soft" X-rays (0.5–2 keV); *green,* medium X-rays (2–4.5 keV); and *blue:* "hard" X-rays (4.5–10 keV). (Courtesy of the Max Planck Institute for Extraterrestrial Physics.)

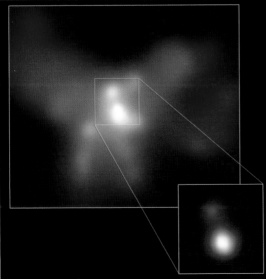

PLATE 26A; 26B. The galaxy NGC 6240 in the middle of a merger process. *Top:* A composite Hubble and Spitzer image in the visible and near-infrared light (Courtesy of Hubble/ Spitzer, NASA), clearly showing tidal tails and dust lanes as well as the two cores of the original galaxies in the center. *Bottom:* An X-ray image taken with Chandra in X-ray colors similar to Plate 25. The red and greenish emission regions show hot gas from exploded stars in the center of the merger event while the two blue nuclei, zoomed in the cutout (*lower right*) represent two active black holes in the two cores of the original galaxies. (Courtesy of Chandra, S. Komossa, Max Planck Institute for Extraterrestrial Physics.)

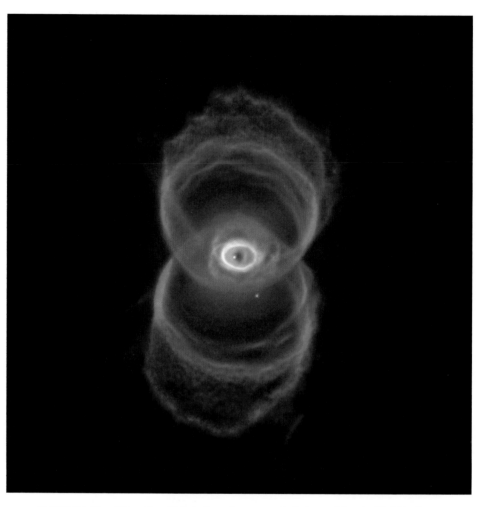

PLATE 27. The "Hourglass Nebula," a planetary nebula imaged by the Hubble Space Telescope—and a model for the future of our sun. (Courtesy of Raghvendra Sahai and John Trauger, Jet Propulsion Laboratory; the Wide-Field Planetary Camera 2 science team; and NASA.)

year later, which has already happened, but no massive outburst has been observed. Probably, the gas stream just added to the total mass swirling around the black hole.[7]

Could an entire star get too close to the black hole and be swallowed? Maybe even Genzel's famous S2? In fact, stars lose some energy in each cycle around the black hole and thus get closer and closer to such a fate. But that takes a long time. Rather, it can happen that a star, disturbed by the close passage of another star, goes off track. Depending on which path it takes when approaching the black hole, it is either ejected from the center at high speed or swallowed. In the Galactic Center, this process is expected to take place approximately every ten to a hundred thousand years. We will see, further below, what happens then.

A Black Hole in Every Galaxy

Astrophysicists have long suspected that a black hole lies at the center of our neighboring galaxy, the Andromeda Nebula (M31). In the late 1980s, spectroscopic methods revealed anomalously large stellar velocities in the center, which pointed to a compact central object of about fifty million solar masses. Later, in the early 1990s, the Hubble Space Telescope identified two compact knots at the center of M31. Initially, these were believed to be two nuclei, each with a black hole, leaving the exact location of the center of our neighboring galaxy uncertain. Such a configuration with two cores would be unstable; the nuclei would have to merge with each other in a relatively short time. The original idea was rejected when it was shown that one of the two knots is probably a ring of old, red stars that orbit the central nucleus.

Spectroscopic measurements with the Hubble Space Telescope, carried out under the leadership of my colleague Ralf Bender, show a mysterious bluish light emanating from the center of our neighboring galaxy. An amazing collection of hot, young stars orbit the center of M31 in a disk with a diameter of about one light-year that is embedded in the ring of red stars mentioned previously. The approximately four hundred stars in the blue disk must have formed about 200 million years ago in a massive star-forming eruption. Similar to the young stars in the Galactic Center, how such stars can arise in the vicinity of the central object remains a mystery. In such a hostile environment, tidal forces should immediately shred all gas clouds from which stars might form. Nevertheless, the blue, young stars in the two galactic nuclei closest to us indicate that this still unexplained phenomenon is probably quite normal. Bender's group also determined the velocities of the blue stars in the center of the Andromeda Nebula using the Doppler effect: on average, they speed around the central object at about one thousand kilometers per second, making the putative black hole 140 million solar masses. Just as in the center of our Milky Way, a black hole is now the most plausible explanation for the compact object in M31.

There exists a third galaxy, in the center of which we can prove the existence of a supermassive black hole beyond any doubt. The active galaxy NGC 4258 contains a central black hole that is currently efficiently fed and therefore emits light in all wavelengths. The accretion disk already mentioned, in which the matter flows inward to the gravity monster, is oriented in a special way. The radio emission from the central object is greatly amplified by the water maser effect[8] through matter nodes in the disk. With the excellent resolving power of radio telescopes, both the location and the rotation velocity of these nodes can be determined with high precision, and consequently, the Kepler orbits of the matter in the accretion disk. As in the case of single stars, one can again infer the mass and the size of the central massive object with high accuracy. Once again, a black hole is the only plausible explanation for such a compact structure. In this case, the very luminous electromagnetic radiation from the bright center of the disk is an additional indicator of a black hole.

In the previous sections, we learned about the three best-defined black holes in the centers of nearby galaxies and also the different methods for measuring the mass of a black hole. In all cases, we observed how the gravity monster affects the matter in its environment: in the Galactic Center, we saw individual stars thrown around at high speeds; in the Andromeda Nebula, we observed the high velocity of the cluster of blue stars; and in NGC 4258, we directly witnessed the gas orbiting the black hole in an accretion disk. Similarly, but using a much lower spatial resolution, scientists have determined the masses of seventy-two black holes in the centers of nearby galaxies in recent years. Once more, it turns out that our galaxy is nothing special. The most massive compact objects contain up to three billion solar masses, a thousand times more than the one in our Milky Way. Interestingly, a direct correlation exists between the mass of the black hole and the size of its parent galaxy. If we plot the mass of the black hole against two variables that, in essence, depend on the size of the galaxy—the luminosity of the central bulge, on one hand, and the velocity at which the stars in the center of the galaxy move, on the other hand, we find surprisingly good correlations.[9]

So there must be a connection here that has been hitherto not understood. The central black hole and its host galaxy seem to mutually "know" each other, but the black hole does not in any way dominate its host galaxy. On average, its mass is only about 0.2 percent of the mass of all the stars in the galaxy. The black hole therefore seems to be an integral part of the history of the galaxy and, from the beginning, seems to grow hand-in-hand with it.

Some time ago, Stefanie Komossa and her colleagues in our group at the MPE, Garching, found another piece in the puzzle of central supermassive black holes. The current British Astronomer Royal, Martin Rees, who is also Baron Rees of Ludlow, predicted in the 1980s that individual stars that get too close to a central galactic black hole will be captured and ripped apart by its enormous tidal forces. The black hole will eventually swallow the remnants of the star, which should lead to a dramatic luminosity increase at the center of the galaxy and an X-ray "afterglow" lasting for several years. In Martin Rees' view, this would be the best proof for the existence of a central supermassive black hole in a galaxy. Some scientists use the fact that such events have never been observed to argue against the existence of black holes. Although in each individual galaxy the tidal capture and disruption of a star is rather rare—it should take place every ten thousand to one hundred thousand years—this should happen frequently across the whole sky. Indeed, the ROSAT mission found a small number of events in which an apparently perfectly normal galaxy temporarily became dramatically brighter. It was immediately speculated that these could be such tidal disruption events; however, other interpretations could not be ruled out. Komossa and her colleagues observed one of these sources with Chandra and XMM-Newton about ten years after the outburst. They found a very weak X-ray source having the characteristic spectrum and luminosity of an accreting, supermassive black hole exactly in the center of this galaxy, thus confirming Martin Rees' prediction. In the meantime, using data from the Galaxy Evolution Explorer (GALEX) space mission observing in the ultraviolet range and the Panoramic Survey Telescope and Rapid Response System (Pan-STARRS) observatory, Suvi Gezari and her coauthors identified the first tidal capture event while it was still going on, including the all-important rising phase of the outburst.[10] For future X-ray surveys, like the one planned with the eROSITA (extended ROentgen Survey with an Imaging Telescope Array) instrument, we hope to catch many more such dramatic events in the act and study them in detail.

The Resolution of the X-ray Background

The astrophysicist Riccardo Giacconi can look back on a fascinating career as a scientist and science organizer. In 2002, he was awarded the Nobel Prize in Physics "for pioneering contributions to astrophysics, which have led to the discovery of cosmic X-ray sources." I had the pleasure of working with him and learning from him intensively for more than a decade. Giacconi was born in Genoa in 1931 and studied physics at the University of Milan. He earned a doctorate there and immediately became assistant professor of physics. At the age of twenty-five, he traveled to the United States, first to the University of Indiana in Bloomington, and later to Princeton University. His career as a pioneer of space exploration began after receiving an offer from American Science & Engineering Inc. (AS&E) in Cambridge, Massachusetts. This company, founded by Chief Executive Officer Bruno Rossi, performed research and development for the government. When Giacconi joined the company in 1959, AS&E had twenty-eight employees. Right after the Sputnik shock, he was charged with building up a space program for the company and given great freedom for research. Judging from Giacconi's autobiography,[11] the years 1959 through 1962 were among the most productive of his life. He "participated in classified research, a total of 19 rocket payloads, six satellite payloads, an entire satellite and an aircraft payload, and four rocket payloads for geophysical research." In these two and a half years, Giacconi's group within the company grew from the original three to more than seventy employees.

The year 1960 saw the beginnings of X-ray astronomy. At a party at his house, Rossi reported on discussions at the National Academy of Sciences regarding the potential of X-ray astronomy and suggested that the company invest in this area. Immediately thereafter, Giacconi constructed a report on the theoretical and experimental possibilities of X-ray astronomy.[12] X-rays are absorbed by Earth's atmosphere, so instruments for their observation must be taken into space. In 1949, US physicists, along with Herbert Friedman of the Naval Research Laboratory, launched a converted V-2 rocket left over from World War II and discovered the X-ray radiation from the sun, the brightest

X-ray source in our vicinity. In their first estimate, Giacconi and his colleagues concluded that their existing instruments would never be able to observe the X-ray emissions from stars radiating as brightly as the sun due to their much greater distance. The same report also discussed supernova remnants and other peculiar celestial sources with huge uncertainties regarding their possible X-ray emission. They concluded that only by observing the moon, whose surface reflects the solar X-rays, would they be certain to make discoveries. Around this time, NASA was planning the Apollo Program, and Giacconi convinced the government agencies to provide funds for developing a rocket payload fifty to one hundred times more sensitive than any instruments ever used in order to observe the moon's fluorescence radiation. They hoped to study the chemical composition on the lunar surface long before an astronaut set foot on the moon.

On June 12, 1962, everything was finally ready. Two previous rocket launches had, unfortunately, been unsuccessful. But this time, the rocket rose successfully and once outside the Earth's atmosphere, scanned the whole sky for about five minutes. The highly sensitive detectors worked wonderfully and recorded the intensity from different directions. The results proved fundamental in two respects.[13] First, the scientists around Giacconi discovered the signal of a very strong X-ray source in the sky—much stronger than anything they had expected. Although the signal came roughly from the direction of the moon, it was almost thirty degrees away from it and therefore clearly not connected with Earth's companion. Because they located the source in the Scorpius constellation, they named it Scorpius X-1 (or abbreviated Sco X-1), the first X-ray source in this constellation. Years later, ever more precise measurements identified this source as an X-ray binary star system, in which a neutron star accretes matter from a normal companion star. In contrast to a "normal" star like the sun, in which X-ray light comprises only about one millionth of its total radiated energy, Sco X-1 is an X-star, which behaves in the opposite manner: it emits the majority of its radiation via X-ray light. Over the years, many other bright X-ray stars have been discovered in other constellations and named accordingly: Cygnus X-1, Cygnus X-2, Hercules X-1, etc. Many years later, in 1989, I worked on the detailed spectral and time variability analysis of this class of X-ray binary star sources and published a seminal work,[14] which is still my most-cited paper. A perhaps even bigger surprise, however, was the first successful rocket flight's discovery of a constant background radiation across the whole sky. Taken together, this diffuse radiation is about as bright as Sco X-1, the brightest discrete source in the sky. If we had X-ray eyes (and could be outside Earth's atmosphere), we would need no reading light at night because the sky would shine brightly enough. The diffuse X-ray glow was the first extragalactic background radiation discovered—a few years before the microwave background radiation discussed in Chapter 3.

Together with Bruno Rossi, Giacconi also developed the concept of an imaging X-ray telescope in 1960, forming the basis on which all later X-ray optics operated in space. He gleaned ideas from German physicist Hans Wolter, who originally wanted to use these optics for X-ray microscopy in 1954. In the title of the 1960 paper published in a geophysical scientific journal,[15] the word "telescope" is, interestingly, still set in quotation marks. In 1963, Giacconi, along with Herb Gursky, created a road map for the future development of X-ray astronomy. It began with additional rocket experiments followed by satellites. First, simple, nonimaging X-ray detectors should be used, then the first imaging X-ray telescope, and finally, a large X-ray satellite with a 1.2-meter mirror system to resolve the X-ray background into individual discrete sources. Giacconi and Gursky originally believed this plan could be realized within five years—but that proved far too optimistic. The process took until the end of the millennium, when all components of the original plan were eventually realized.

Uhuru was the first satellite developed under Giacconi's leadership. It was launched in December 1970 from a floating platform off the coast of Kenya. The name *Uhuru,* which means "freedom" in Swahili, was chosen to honor the people of Kenya because the launch fell exactly on the seventh anniversary of this African country's independence. Uhuru discovered about four hundred X-ray sources in the entire sky, most of them compact X-ray binary systems in the plane of the Milky Way but also a number of extragalactic sources. The second satellite built under Riccardo Giacconi's direction was renamed the Einstein Observatory following its successful launch in November 1978. It carried the first nonsolar imaging X-ray telescope with four nested mirror shells in the shape prescribed by Hans Wolter: a paraboloid connected to a hyperboloid. The largest shell had a diameter of fifty-eight centimeters, and the telescope had a resolution of about five arc seconds. The mission was completed in April 1981. Giacconi's biggest dream was realized only in July 1999. As one of the NASA "Great Observatories," Chandra is now one of the main workhorses of X-ray astronomy, together with ESA's XMM-Newton. Chandra, with an image quality of 0.5 arc seconds, is the highest angular resolution X-ray telescope ever built and flown. Because no major new high-resolution X-ray telescope is planned for the foreseeable future, Chandra will hold this honor for decades to come.

In parallel, Joachim Trümper furthered the field of X-ray astronomy in Germany. Like several other international high-energy astrophysicists, he had started his career with the study of cosmic rays. Around 1970, he began working in the new science of X-ray astronomy, in particular on the physics of neutron stars, a field in which he had and still has great achievements. After he moved from the University of Kiel to the University of Tübingen in the early 1970s, he built an instrumentation and balloon flight program in X-ray astronomy there. In 1975, Trümper was appointed as a member of the Max Planck Society, where he took up the position of director at the MPE in Garching, Germany. The most important result of the balloon payload HEXE (High-Energy X-ray Experiment), jointly developed by the MPE and the University of Tübingen, was the discovery of a cyclotron line in the spectrum of the bright X-ray pulsar Hercules X-1, allowing the first direct measurement of the gigantic magnetic field on the surface of a neutron. The magnetic field of Her X-1 is about $5 \cdot 10^{12}$ gauss, approximately five quadrillion times higher than Earth's magnetic field.[16] In 1972, together with his colleagues in Tübingen, Trümper began to investigate observation techniques with the aid of imaging X-ray optics. In 1974, he submitted the proposal for a national X-ray satellite to the German Federal Ministry of Research, which years later turned into ROSAT. Together with the Carl Zeiss company in Oberkochen, his team developed small thirty-two-centimeter Wolter telescopes, which were launched on Skylark rockets along with very sensitive imaging proportional counters developed at MPE. The first of three successful rocket flights launched in spring 1979 from the Australian base Woomera. It took the first X-ray color images of a cosmic X-ray source (the supernova remnant Puppis A) a few weeks before the NASA Einstein Observatory began its operations.

At this point, I need to weave in some of my own biography. My enthusiasm for physics and astrophysics actually came only at the end of high school and while studying at Munich University. In school, I was fascinated by books about modern science, for example, Werner Heisenberg's philosophical notes *Schritte über grenzen* (Steps beyond boundaries),[17] and I devoured Hoimar von Ditfurth's evolutionary book *Im Anfang war der Wasserstoff* (In the beginning was hydro-gen),[18] as did my current wife, Barbara. But the decision to study physics and later even astronomy resulted from many detours—among others my career as a rock musician in the band Saffran. I found my real passion for astronomy through Rudolf Kippenhahn's astronomi-cal lectures, which he published shortly afterward under the title *100 Billion Suns: The Birth, Life and Death of the Stars*.[19] He held these excellent lectures at the Ludwig Maximilians University for students of all faculties and always had a full lecture hall. They were very pop-ular because Kippenhahn had the gift of making rather complicated subjects easy to understand. In summer 1978, I had the good fortune to obtain my first astronomical observing experience at the University Observatory in Munich–Bogenhausen. On every clear night through-out the whole summer, I took photographic spectra of a bright nova that had just erupted in the Cygnus constellation. As expected, the nova Cygni 1978 became progressively weaker in the weeks and months after the explosion, forcing me to upgrade the telescope's ac-quisition and guiding system to continue observing this phenomenon. The whole thing came abruptly to an end when Oktoberfest illumi-nated the sky over Munich so brightly that you could not see stars anymore. In winter 1978 and 1979, I attended Joachim Trümper's fas-cinating lectures on X-ray astronomy. I still remember one afternoon in the Schellingstrasse, where he challenged us to perhaps one day solve the mystery of the asymmetric pulses of the rapidly rotating neutron star in the Crab Nebula.

Toward the end of the semester, I asked him for advice in choos-ing a diploma thesis. He said, "Why don't you just come with me," ushered me into his car, and drove me out to the MPE in Garching, where I have spent a substantial fraction of my scientific career. In spring 1979, while a large part of the MPE team participated in the aforementioned rocket launch campaign in Australia, I started my di-ploma thesis on the scattering of X-rays on polished surfaces. This was to prepare for the ROSAT mission, which required extremely well-polished X-ray mirrors. The Wolter telescopes are irradiated by X-rays at very slanted angles and reflected twice. Because X-rays have a very short wavelength, they see the tiniest bumps on the mirror sur-face as gigantic mountains, unlike the situation with visible light, with a typical wavelength one thousand times longer. Therefore, X-ray mir-rors must be so exquisitely well polished that the average amplitude of the surface irregularities is only a few atomic layers. At that time, this presented an unsolved problem. During my thesis, the Carl Zeiss company, in collaboration with the MPE, succeeded in developing ever more accurately polished mirror samples. We measured the surface roughness of these samples at the MPE's vacuum X-ray test facilities and communicated the results directly back to the polishing specialists at Zeiss. This work later led to the ROSAT mirrors' entry into *The Guinness Book of World Records* as the smoothest surfaces in the world.

In the meantime, scientists of the X-ray groups in Garching and Tübingen, who had originally discovered the cyclotron line in Hercules X-1, had carried out several more successful balloon flights and prepared a new, much larger balloon gondola that was supposed to carry three HEXE detectors and a new high-resolution germanium semiconductor detector. At that time, the brand new CCD (Charge Coupled Device) detectors, which nowadays are built into every digital camera and smartphone, became commercially available. As part of my thesis project, I was tasked with developing a separate CCD star camera for the balloon gondola. I had a lot of fun designing the camera, soldering its electronics together, and later programming its chips. I was also responsible for maintaining and calibrating the semiconductor detector, which had to be continuously cooled with liquid nitrogen. This work culminated in two balloon campaigns: in 1981 to Palestine, Texas, and in 1982 to Uberaba, Brazil. The high-altitude research balloons can carry payloads of over a metric ton into the stratosphere, the uppermost layer of Earth's atmosphere at forty kilometers altitude. On the ground, a small part of the balloon is filled with helium (sometimes also with hydrogen, but this is very dangerous). The balloon first rises like a huge elongated sausage, puffing up as it reaches the thinner layers of the atmosphere until it floats in the stratosphere like a ball about one hundred meters in diameter. Once there, the high winds of the jet stream blow it away. Twice a year, however, the western wind direction turns around for a short time. These so-called "turnarounds" see a pilgrimage of high-altitude balloon researchers from all nations trying to launch their payloads into the air. Because the average wind speeds during the turnarounds are much lower, high-altitude research balloons, with some luck, can stay up in the air for several days sending data. However, during a launch in the low layers of the atmosphere, absolutely no wind is allowed; otherwise, the extremely thin balloon skin is in danger of being damaged during the release. Because of the always unpredictable vagaries of the weather and the material defects that occurred quite frequently at that time, balloon flying was a very risky business. Of the two flights I was involved in, the first was a great success, but the second had to be aborted after a few hours because of a balloon failure. Nevertheless, we obtained nice data regarding the Crab pulsar, and as predicted by Joachim Trümper, I could indeed address the challenge of its asymmetric pulse structure in my PhD thesis.

In the meantime, the German Federal Research Ministry approved the ROSAT mission, and when I finished with my thesis, I was immediately hired into the ROSAT team. There I had several tasks. Because of my experience with polished surfaces, I was involved in the calibration of the ROSAT mirror system but could also participate in the calibration of the ROSAT focal plane detectors. The imaging proportional counters (PSPC: Position Sensitive Proportional Counter), which a team around our colleague Elmar Pfeffermann developed, belonged to the outstanding technological achievements of their time and made a major contribution to ROSAT's overall success. Even today, the large MPE X-ray vacuum test facility PANTER in Munich-Neuried uses a spare model of such a detector. In the years before the ROSAT launch, this test facility regularly performed calibration measurements for the mirrors and the detectors. My small group had the task of immediately analyzing the data tapes that arrived in the evenings from Neuried; if possible, on the same night, to give the calibration team a rough feedback for the measurements to be performed the next day.

Perhaps due to my experience with the HEXE balloon gondola's star camera, I was given the great scientific responsibility of supervising ROSAT's attitude measurement and control system. We sent the calibration measurements of the star sensors and the so-called hardware-in-the-loop test to the company MBB (Messerschmidt, Bölkow & Blohm) in Ottobrunn near Munich, which performed a realistic end-to-end simulation of the whole satellite attitude control system. These activities would later turn out to be particularly important because one of ROSAT's four gyroscopes failed shortly after the beginning of the mission and endangered the entire project. In a cooperation between industry, scientists, and the German Space Operations (GSOC) in Oberpfaffenhofen, we had to reprogram the gyro control of the satellite into a magnetic compass control, practically like an open-heart surgery. The concept of magnetic field control for three-axis stabilized satellites, which was brand new at that time, prolonged the active life of ROSAT for many years, even after several other gyro failures, and has now become standard in a number of satellite systems.

A realistic computer simulation of the mission proved to be extremely important and later contributed to the immediate understanding and the rapid scientific utilization of the ROSAT data. Starting with idealized assumptions but later fed with increasing volumes of detailed information from the calibration of the instruments, we simulated observations of the ROSAT mission as closely as possible. Every single X-ray photon was artificially created in the sky and then tracked on its way through the mirror system and the detector. Over time, these simulations became more and more realistic. We included point-like and extended objects in the sky, different spectral models, and even the temporal variability of the X-ray emission. We also simulated the background radiation, the extragalactic cosmic ray radiation, the solar scattered radiation, and the intrinsic background of the detector as realistically as possible. This gave us access to fairly realistic data long before the real mission, and we were even able to modify and improve the satellite's flight software.

The support structure for the PSPC cameras created one concern. These gas proportional counters had a very thin entrance window for the X-ray photons. The pressure of the counter gas would burst the window if it were not protected by a highly complicated and finely chiseled network of wires. However, this network cast a shadow on each X-ray image that the satellite received. At the beginning, we hoped that the inherent jitter of the ROSAT attitude control system would wash out these shadows. However, when we fed the results of the first realistic attitude control tests into our simulation program, it soon became clear that the satellite orientation was much too stable. This happening just before the completion of the mission preparation did not make it easy to explain to the engineers—especially the main contractor, Dornier System in Friedrichshafen, as well as the funding agency in Bonn—that the system had been constructed too well and thus we needed to include an artificial dithering motion in the flight software. I convinced them with the argument that even the human eye has a blind spot, compensated for by the constant movement of the eyeball and the corresponding data analysis in the brain. This "wobble mode," in which the viewing direction of the satellite wandered back and forward on a time scale of a few minutes, later contributed significantly to the high quality of the ROSAT observations and was even implemented in the NASA Chandra observatory in an expanded form.

In 1985, Joachim Trümper assembled an international team of
scientists in Garching who were all interested in performing very long
and sensitive surveys of the X-ray background radiation. Riccardo Giacconi, who controlled the fate of the Hubble Space Telescope at the
Space Telescope Science Institute in Baltimore, brought along his young
colleague Richard Burg. The two had experience with the deep Einstein
surveys and proposed to focus on a sky area that we later called the
Lockman Hole. This area, where the density of the Milky Way shows
an absolute minimum and therefore represents an ideal window into
the extragalactic universe, had just been discovered by US radio astronomer Jay Lockman and his colleagues. Gianni Zamorani from the Astronomical Observatory in Bologna had already gained a lot of experience with observing faint quasars in visible light. Maarten Schmidt from
the California Institute of Technology, the discoverer of quasars, had
previously calculated detailed predictions of how many X-ray sources
should be detected in deep surveys with ROSAT. He planned the optical
follow-up observations, first with the big five-meter mirror on Mount
Palomar and later with the ten-meter Keck telescopes in Hawai'i.

For more than a decade, this team met every six months for a few
days in Garching and worked on the preparation and implementation
of the deep ROSAT surveys. During the first meeting I, as a young
buck, was entrusted with the responsibility for the "Deep Survey" project. It was a challenge but also a great opportunity. We dealt with
astrophysical backgrounds; detailed simulations of the observations;
the tricks of data analysis—in particular, their statistical and systematic effects; the preparation and implementation of the optical observation campaigns; writing applications for the ROSAT observing time;
and later, with the analysis of the first results. I still well remember
many of our meetings. Riccardo and Maarten, although very construc-
tive, were usually also very critical. Sometimes their opinions differed,
and we argued for hours with great verve. But often they also disagreed
with my proposals for the observational strategy or the data analysis
methods. In these cases, I had to spend most of the night with new
simulations in order to convince them the next morning. Riccardo
once said that the ROSAT mission was far better prepared and the
data analysis better understood before the launch than was the case for
the Einstein mission even years after its completion. I regarded this as
a great compliment. The highlight of these meetings was usually a
sumptuous dinner at the Augustiner Bräu in Munich, where Riccardo
particularly appreciated the excellent quality of the *Apfelstrudel*.

The Lockman Hole actually covers several degrees over the sky. At first, we had to determine the optimal location for our survey. At some point, Maarten Schmidt realized that the field was very close to one of the bright stars of the Big Dipper, which we were able to avoid at the last minute before the start of the observations. We started an optical survey of the field well before the ROSAT launch. Later, many other groups selected this area for deep surveys in other wavelength regions. The Lockman Hole became one of the best-studied extragalactic fields in the sky.

I have already reported on the dramatic night of the *Challenger* catastrophe, which delayed the ROSAT mission for several years. After intense preparations and a long waiting time, ROSAT was finally launched successfully on June 1, 1990, on a Delta rocket from Cape Canaveral in Florida. A particularly exciting moment was the important "first light," when the PSPC detector was initially switched on after the telescope opened its eyes. On June 16, a large crowd from the ROSAT team and some guests gathered at the GSOC in Oberpfaffenhofen. Of course, we were all curious to see if the X-rays would really be imaged correctly in the telescope. I myself was particularly nervous because this was the first time the wobble mode had been tried in real life. We picked a date on which we had direct contact with the satellite from a ground station in Australia and a location targeting the Large Magellanic Cloud, where supernova 1987A had exploded. Everyone held their breath as the first single X-ray photons appeared on the screen. As the image of the Magellanic Cloud gradually assumed contours, cheers and standing ovations broke out. Embarrassingly, however, I had made a mistake when initially programming the attitude software—either a wrong sign or a reversed trigonometric function. The wobble mode of the satellite was not properly corrected out but on the contrary, was amplified so that all detected X-ray sources were drawn out into elongated traces on the sky. However, this did not detract from the general joy of ROSAT successfully opening its eyes.

The first weeks and months of the mission proved very dramatic and exhausting, but the intense preparation from the entire team was a great help. Together with my colleague Peter Predehl, we developed a first preliminary data analysis system from the calibration and simulation software, which we called Trampelpfad ("trodden trail"), through which we were able to supply researchers worldwide with ROSAT data. As one of the first spectacular ROSAT observations, we repeated Giacconi's original experiment twenty-eight years after the

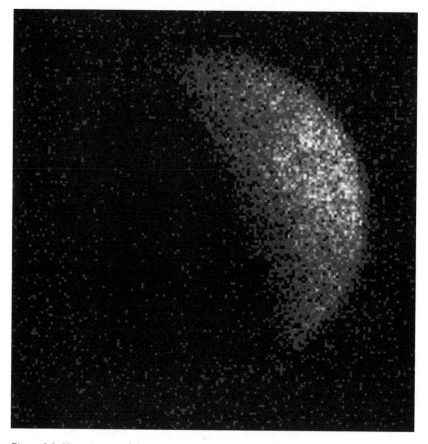

Figure 8.2. X-ray image of the moon with the Roentgen Satellite (ROSAT) Position Sensitive Proportional Counter (PSPC). Each individual point represents a measured X-ray photon. The crescent of the moon shows the X-rays reflected by the sun. The dark side of the moon casts a clear shadow on the cosmic X-ray background radiation. (Courtesy of the Max Planck Institute for Extraterrestrial Physics.)

first rocket flight and observed the moon in X-rays. The X-ray image of the moon (Figure 8.2)[20] is particularly fascinating because the dark side of the moon really casts a shadow on the diffuse X-rays behind it. With this picture, it became clear that ROSAT's extremely low intrinsic background signal made it perfectly suited for the study of diffuse X-ray radiation.

In the first six months of its mission, ROSAT performed the first all-sky survey with an X-ray telescope. Comparing the ROSAT map of the X-ray sky (Plate 24) with the image of the Milky Way in visible light (Figure 1.1), the regions of the galactic plane that are bright in visible light appear darker in X-rays because the diffuse emission in the background is absorbed. The areas above the galactic plane, which are dark in visible light, appear in brightly colored X-ray light. These are the explosion clouds of dead stars that cover the entire sky like a fog. The extragalactic background radiation discovered by Riccardo Giacconi and his colleagues is visible only at higher energies. The Lockman Hole is located at the top left in the reddish area of the sky.

The detailed ROSAT deep surveys in the Lockman Hole began in spring 1991 with a PSPC exposure lasting about two days. A year later, a second equally long PSPC exposure followed. In addition to the two MPE imaging proportional counters, which measured X-ray colors, the satellite also carried a High-Resolution Imager (HRI) X-ray camera, developed for NASA by Steve Murray at the Smithsonian Astrophysical Observatory. This became the workhorse for the satellite after the proportional counter gas ran out. With this camera, we succeeded for the first time in getting a "megasecond" exposure— that is, one million seconds of observations in the Lockman Hole. All in all, ROSAT has observed this field for about 1.4 million seconds (see Plate 25). Considering that the satellite could effectively use only about half the total time because of the Earth's occultations, this corresponds to nearly an entire month of observation time. This is quite comparable with the most sensitive observations in other wavelength ranges, such as the famous Hubble Ultra Deep Field (see Plate 6), and has set the standard for future deep X-ray observations with XMM-Newton and Chandra. First with ROSAT and later with Chandra and XMM-Newton, Giacconi's original dream has thus come true: the resolution of the X-ray background into discrete objects.[21] With these deep surveys in X-ray light, we have performed a feat similar to Galileo Galilei's when he pointed his telescope toward the Milky Way in 1609 and found that it consisted of millions of individual stars. The diffuse X-ray glow in the sky has been resolved with the deepest X-ray observations into hundreds of millions of individual points of light across the whole sky.

Thereafter, the most tedious work consisted of obtaining optical follow-up and spectroscopic identification as well as classifying the newly discovered X-ray sources. Taking spectra of hundreds of weak sources using large optical telescopes proved a Sisyphean task. With the help of these spectra, X-ray sources had to be identified and classified in order to determine their redshift and thus correspondingly, their age or distance. The objects discovered in the deep surveys represented the faintest X-ray sources ever seen and were typically very weak in visible light. To take optical spectra of those, we needed access to the largest telescopes in the world. Maarten Schmidt introduced me to such a world. Together with Don Schneider and Jim Gunn, he developed the "four-shooter," a special spectrographic camera for the five-meter telescope on Mount Palomar. I was invited to attend the first observing run with this giant eye. Maarten, Don, and Jim, three pioneers in the search for ever more distant quasars, are the main characters in Richard Preston's interesting biographical novel *First Light: The Search for the Edge of the Universe*,[22] which portrays the lives of the astronomers on "the mountain" in great detail.

On my first day at Mount Palomar, Maarten showed me the venerable five-meter telescope. In a specially designed elevator, we went up to the prime focus cabin located more than twenty feet above the main mirror. In earlier times, astronomers had to squeeze into this tiny cabin for their observations, often throughout the whole night. Maarten is probably the astronomer who had been in this cage for the longest period of time. Afterward, we climbed through the scaffolding of the dome interior out through a small hatch onto the roof of this gigantic cathedral—on which only the technicians were allowed for clearing snow. This proved to be an absolutely spectacular introduction into optical astronomy!

After the first ten-meter Keck telescope had started its operations, Maarten and I traveled to Hawai'i for several consecutive years every spring to carry out observations in the Lockman Hole. Usually, these astronomers are envied because they carry out their work in such a beautiful environment. However, the trip to the top of the Maunakea volcano at a 4,300-meter altitude on the big island of Hawai'i is very long and arduous, particularly when you come from the other side of Earth. You make multiple stops, with long layovers at airports, and must sleep an additional night at the Hale Pohaku Lodge halfway up the mountain in order to acclimatize. Unlike the tourists in shorts and slippers usually crammed into the plane, the astronomers are heavily loaded with work material and winter gear in the face of extreme cold on the mountain. Working at night at 4,300 meters is no picnic. Every movement is difficult and, because of the lack of oxygen, can lead quickly to a stabbing headache. Your brain and eyesight fail to work at full capacity. Maarten had therefore forbidden us to carry out complicated calculations or to make important short-term decisions in this environment. Every iota of the observing plan had to be prepared at lower levels. Nevertheless, the observing nights at the Keck telescope on the summit of Maunakea (and later on ESO's Very Large Telescope on the summit of Cerro Paranal in Chile) are my fondest professional memories. This was when I started to fall in love with Hawai'i.

Later, the Keck Observatory moved more toward remote observing. The astronomers work at the headquarters building, Waimea, at the base of Maunakea and are connected to the night astronomer and the technicians on the summit via a microwave radio and video link. Maarten struggled for quite some time against this—in his opinion, degenerate—way of doing astronomy. He wanted to hear and smell the telescope, periodically check the sky with his own eyes, and especially, personally communicate with the night astronomer operating the telescope. In fact, in Waimea it is rather common to experience thick clouds and rain while the most wonderful weather prevails at the summit. It is a strange feeling to travel halfway around the world and then stay a few kilometers away from the target of your desire to carry out observations under a blanket of clouds. But in the meantime, practically all observers and even most of the night assistants observe remotely, and many do not even travel anymore to Hawai'i.

Back to science: after many years of optical observations, we obtained a complete picture of the sources of the X-ray background. We showed that most of the newly discovered X-ray sources consisted of distant active galaxies containing well-fed and thriving black holes at their centers. The X-ray background radiation therefore stems from the feeding and growth phase of the entire massive black hole population in the universe. All supermassive black holes we observe today in the centers of nearby galaxies must have grown in earlier cosmic times and imprinted the light emitted during their accretion processes onto the X-ray background. By measuring their redshift, we can place sources into the context of their cosmic history. It turns out that the black holes with the highest luminosity, the quasars, originated in a kind of "feeding frenzy" a few billion years after the Big Bang while the less luminous specimens came much later. This behavior is somewhat paradoxical, as in a hierarchically growing universe one would expect that small structures form first and larger ones grow later. Only in recent years do we understand that this "downsizing" effect came about because in the early universe, galaxies collided with each other much more frequently. As we will see in the next section, this process can feed black holes very efficiently, while later in the universe, the black holes starve over long periods of time.

Galaxy Cannibalism, or The Wedding of Giants

Earlier, we discussed the discovery that the central black hole in our own Milky Way is in principle much too dim, given the amount of gas available in its surroundings. We have come to realize the important role that angular momentum plays during the accretion process. The matter flowing toward the black hole forms a thin disk stabilized by the angular momentum and does not even begin to spiral down into the black hole—as in the case of the rings of Saturn. However, if this accretion disk achieves a sufficiently high density and viscosity, the friction of matter dominates and heats up the disk. The matter is slowly transported inward, efficiently feeding the black hole. This process may play an important role when two galaxies collide, and their central black holes presumably also merge.

The observations of the irregular galaxy NGC 6240 show two galaxies that are currently undergoing a merging process (see Plate 26a). Some astronomers call this "galaxy cannibalism" because it looks as if the two galaxies are swallowing each other. However, at the end of the merger, a larger and perhaps more beautiful galaxy forms, so I prefer to call this process a "galaxy marriage." Simulations of such collisions show two spiral galaxies that cut through each other like rotating saw blades and twist around the orbits of their stars. In this case, the angular momentum of the galaxies is significantly redistributed. Large tidal tails of stars and gas are pulled out of the two galaxies and flung far into space. At the same time, stars and gas masses swirl into the two galaxy centers and cause violent bursts of star formation. In simulations, it takes several hundred million years until both galaxies have completely merged and a new, elliptical galaxy develops. It has long been believed that in this merging process, the two central black holes of the original galaxies also collide, accompanied by a strong outbreak of gravitational radiation, and thereafter form the powerful quasars that outshine their galaxies in all wavebands.

The X-ray observation of NGC 6240 was the first discovery in which two active black holes were caught in a single galaxy "in flagrante delicto" (see Plate 26b). This proved to be the key to understanding several unanswered questions of galaxy evolution. Apparently, nature succeeds in feeding the two black holes so efficiently that they grow very rapidly and shine brightly long before they actually collide and merge. Although they cannot be detected in visible light because they hide behind dense gas and dust clouds, both X-ray nuclei in the galaxy NGC 6240 are as luminous as quasars. From observing other quasars and active galaxies, we know that these can blow away almost as much matter as they swallow. One can imagine this as a strong wind emanating from the quasar, often connected by finely collimated jets of matter hurling particles away from the center of the galaxy at nearly the speed of light (see Plate 23). During the merger process, which lasts hundreds of millions of years, the growing black holes release so much energy that all the free gas masses of the two original spiral galaxies blow outward in the winds—far into intergalactic space. Thus, the newly created elliptical galaxy no longer contains enough fresh cold gas to form new stars. The two young, blue spiral galaxies therefore turn into an old, red, and "dead" elliptical galaxy. The dramatic feedback effects of the central black holes can explain several phenomena previously not understood: the strong correlation between the size of the central black hole and the mass and the stellar velocity of the host galaxy; the fact that elliptical galaxies do not contain cold gas; and finally, the upper mass limit of the galaxies. The central black holes simply do not allow their host galaxies to grow beyond a certain limit and therefore dramatically influence cosmic evolution.

From the redshift of the galaxy NGC 6240, we know that it is located about 400 million light-years from us. We therefore do not see the galaxy as it looks today but see it 400 million years younger due to the distance the light has traveled. The simulations tell us that it should be 100 million years until the two black holes in galaxy NGC 6240 unite. But this means that the event has already taken place, and the gravitational waves—ripples in space-time generated by the two black holes swirling around each other—must already be headed toward us! The gravitational wave space observatory LISA (Laser Interferometer Space Antenna), a masterpiece in laser technology and precision metrology prepared by ESA with the help of NASA, will hopefully begin operations in the next fifteen to twenty years. It will certainly have ceased to operate by the time the gravitational waves from NGC 6240 reach us. In the meantime, several other binary black hole candidates have been discovered. Double black holes like this have gravitational wave astronomers jumping for joy because they provide great hope that such events will occur frequently throughout the observable universe and that LISA will, in fact, be able to measure something.

But what came first: the black hole or the galaxy? Putting together the pieces of the puzzle found in recent years, a reasonably well-rounded picture emerges: about 380,000 years after the hot Big Bang, dark matter dominated the events in the universe. Under its own gravity, it coagulated into an invisible Cosmic Web of filaments and clusters. It dragged the normal matter, from which the galaxies developed, along for this ride. After the Big Bang, the universe waited 200 to 500 million years until it cooled down to the point at which the first stars could form. According to recent theories, the first star in the universe must have originated in what was then the deepest dark matter well (something like a bathtub filled with dark matter and normal matter). This first star must have been very massive, ending its short life in a gigantic supernova explosion. Possibly accompanied by a gamma ray burst, this explosion may have created the first stellar-mass black hole in the universe. Another theory postulates that the first black holes with 10^4 to 10^5 solar masses may have formed in the early universe under very special conditions due to the direct collapse of massive gas clouds. For the rest of their lives, these black holes stayed near some of the largest mass concentrations in the universe and, like the spider in the web, waited until something to eat came along. In a few hundred million years, they expanded to billions of solar masses, thus forming the first distant quasars and simultaneously restricting the further growth of their galaxies. In this image, the first black holes arose as "galaxy seeds"—far ahead of most other stars—and then grew together with their galaxy. The "great feast," the feeding frenzy giving rise to the quasars, can be explained by a particularly large number of galaxy collisions during this "Sturm und Drang" period. At this time in the past, the universe was much smaller, and the galaxies clashed into each other more often. On the other hand, it took some time until the largest galaxies formed. This resulted in a maximum of galaxy marriages at redshift 2–3, only a few billion years after the Big Bang.

The locations where the deepest bathtubs of dark matter resided in the early universe; where the very first stars originated and shortly thereafter may have turned into black holes, contain the richest clusters of galaxies in our universe today. In their centers, these gargantuan galaxies harbor huge black holes that we are still just learning about. The oldest object in our local universe may well be the black hole in galaxy M87 in the center of the Virgo Cluster (see Plate 23).

Chapter Nine

THE DESTINY OF THE UNIVERSE

THE CALENDAR OF THE UNIVERSE

So far, this book has confronted you with so many different size scales, both minute and huge, that I'm afraid you might be losing sight of the big picture. So before we turn our attention to even more inconceivable eternities, I want to slowly bring you back to Earth and to our human dimensions of space and time. Following an idea of the deceased Peter Kafka, a Garching colleague of mine, I am going to demonstrate once again the eventful history of the universe in fast motion. I want to redefine the time scale and squeeze the 13.8 billion years that lie between the Big Bang and today into one single year. The Big Bang occurred on January 1, at 0:00, and the fireworks on New Year's Eve, one year later, represent today. On this scale, one second of our fast-motion year corresponds to 433 years in real time (see Table 2 in the Appendix).

According to this time scale, the matter-radiation decoupling—when the universe possessed the qualities of a candle flame and became transparent—occurs only fourteen minutes after the year begins. Between January 5 and 13—we are not as yet able to determine it any more precisely—the first high-mass stars form, directly followed by the first stellar black holes. Inside their bellies, the stars begin concocting heavy elements up as far as iron and then create even heavier elements during their supernova explosions. From January 20 to 23, the oldest galaxies and quasars that we know of today emit their light. The climax of star formation and galactic cannibalism—the big feeding

175

frenzy—takes place toward the end of March. Then, for many months, only one thing happens: fueled by galaxy collisions and supernova explosions, countless new generations of stars brew more and more heavy elements. This is how, at the beginning of September, the sun, and with it our planetary system, begins to develop. Today, we know that planets, created through the agglomeration of dust, can mainly form around stars with sufficient heavy elements. Therefore, it is not at all surprising that it took such a long time for Earth to come into existence.

At this point, I've got to briefly enter fields that I personally know very little about but nonetheless find fascinating: paleontology and early biology. The first minerals, the zircons, which were found in today's Western Australia, form on September 4, and the earliest coherent rock formations on Earth appear on September 14 in the Canadian Northwest Territories. Chemofossils found in southwestern Greenland in the oldest sedimentary rocks seem to suggest that on September 19, life on Earth already exists. The first traces of organisms in terms of microbe-like, structured filaments and cells, called stromatolites, can already be found by September 29, again in today's Western Australia and in South Africa. Here, we are talking about the cyanobacteria—commonly known as blue-green bacteria or blue-green algae—which are already in a position to reproduce through complex cell division and have the ability to perform photosynthesis.

A great deal of what happened in the long months after cyanobacteria appeared on Earth still remains a mystery. In fact, the cyanobacteria stayed very busy polluting Earth, but in a way, they also "terraformed" it. They poisoned the atmosphere with oxygen—their metabolic end product—leading to a mass extinction. For the following generations, however, this oxygen became essential for life. Around mid-December, all over the world, the first societies of multicellular organisms, the Ediacara fauna, spring to life, including exotic blueprints of long-extinct organisms. Scientists speculate that this new life may have been forced to make several repeat attempts thereafter, perhaps, dying in the global glaciation (which, in our fast-motion year, occurs in mid-December) when the Earth froze over into a giant snowball. Yet new knowledge shows that life is much tougher than so far assumed and even exists several kilometers into the Earth's crust or in hot volcanic springs at the bottom of the ocean. After waiting for many months, this period ends on December 16 in the Cambrian. At this point in time, Earth's continents look totally different. Africa, South

America, and parts of Central America, India, Europe, the Antarctic, and Australia all together form the supercontinent Gondwana. In this "Cambrian explosion," all sorts of new multicellular organisms suddenly develop. The first conchiferous and vertebrate life forms appear and with them the blueprints of today's still existing organisms. Today, many are already extinct again.

Throughout several paleontological periods—such as the Silurian, Devonian, Carboniferous, and Permian—coral reefs, ammonites, and fishes develop in the ocean. The land, which in the meantime has merged into the supercontinent Pangaea, contains horsetails, forest trees, spiders, insects, and reptiles. At this point I always add as a joke: "And the Christmas trees were ready just in time for Christmas." Around Christmas, when the Triassic follows the Permian, the largest of five known mass extinctions in Earth's history occur. It is not entirely clear whether this was due to Pangaea breaking apart. Between 70 and 95 percent of all living organisms die out at this point—but not the dinosaurs or reptiles, which now start their triumphal march. For the next couple of days, they will rule Earth. Roughly on December 25, the first small mammals appear on the scene, as well as, a little later, the "Urvogel" *Archaeopteryx,* possibly a descendant of the small dinosaurs. In a world dominated by dinosaurs, it is rather hard for mammals to evolve.

On December 28, during the transition from the Cretaceous period to the Cenozoic era, the dinosaurs become extinct, as do many other groups of animals, such as the ammonites and other marine organisms. The reason for this mass extinction is a controversial topic among experts. In different places on Earth, iridium-enriched clay has been discovered above the chalk layer, suggesting that a big meteorite impact occurred. A meteorite approximately ten kilometers in diameter created the Chicxulub crater on the edge of the Yucatán Peninsula in Mexico at about the time in question. Many hold it responsible for the extinction of the dinosaurs. But scientists from the United States, Germany, Switzerland, and Mexico have recently analyzed a core sample from a depth of 1,500 meters from the center of the Chicxulub crater that indicates the impact occurred about 300,000 years prior to the end of the Cretaceous period and before the extinction of the dinosaurs. Perhaps another, as yet unidentified, meteorite impact, or combined events, such as increased volcanic activity and multiple meteorite impacts, led to a drastic change in climate and therefore to an extinction of the species.

In the end, the dinosaurs make way for the rapid development of larger mammals. Now the huge mountain ranges on Earth today, the Himalayas and the Alps, surface through the collision of massive continental plates. On December 31, just under four hours to midnight, the hominid line branches off from the rest of the primates. The upright walking man evolves—first the *Australopithecus,* then also other sidelines. The modern human, *Homo sapiens,* starts his existence at only about six minutes to midnight. For a while, he lives together with the *Neanderthal,* the only typical "European" among prehistoric man. However, he represents a less successful sideline that dies out roughly 70 seconds before midnight. Jesus Christ is born about 4.6 seconds before midnight—to be politically correct, one would have to say that all the great world religions developed only during the last 15 seconds of the calendar of the universe. We ourselves—on this scale, a human life of 100 years lasts 0.23 seconds—are only a blink of an eye in the history of the cosmos! Nevertheless, a human life measured in the scale of the cosmos still describes a measurable time unit from our point of view. It could have also been six microseconds (which would correspond to the life span of a mayfly).

We can now risk taking a glance into the future. If we ourselves don't very soon manage to destroy our globe, it will only take until approximately January 12 until Earth will be too hot to live on. As time goes on, the sun will gradually expand, and the water on Earth will begin to boil. On about April 16, the Andromeda Galaxy will swallow the Milky Way (if it doesn't just scrape past after all)—what a spectacular sight for future astronomers! The Galactic Center will then become a quasar, and the two big black holes in the center of the Milky Way and the Andromeda Galaxy will unite with a big roar of gravitational waves. Around July 10, the sun, having used up all its hydrogen, will expand, turn into a red giant, and swallow the inner planets Mercury and Venus. Seen from Earth, the sun will then stretch from horizon to horizon, and the rocks of the inner planets will melt. But with that, our story has not yet come to an end!

THE FUTURE OF THE UNIVERSE: A SPECULATION

After having dealt with the Big Bang; the formation and evolution of the galaxies; the stars; the planets, especially, our Earth; and finally, the end of life, the thread running through this book is now almost fully unwound, and we must turn our attention toward the distant

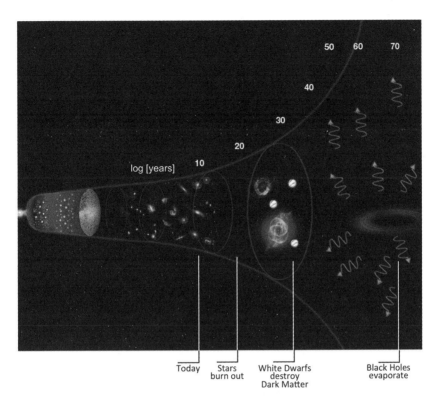

Figure 9.1. Overview of the possible further development of a universe whose expansion is exponentially accelerating. The vertical red lines give the time that has passed since the Big Bang in years to the power of ten (our universe today lies roughly at 10^{10} years). The most important further phases of the universe are indicated. After about 10^{15} years, the last stars will have burned out. After that, the white dwarfs, which are cooling down, will dominate the universe. Around the time of 10^{25} years, dark matter will gradually be destroyed in their interior. After many more eons (e.g., 10^{37-39} years), it is assumed that the protons will decay, and therefore the baryonic matter will be destroyed. For a very long time afterward, the only thing left will be the black holes; they too will evaporate sometime around 10^{70-100} years. In the end, the universe will have thinned out so much that it will vanish in the face of the energy of the vacuum—and "nothing" will be left except, possibly, the same energy field out of which the universe was created to begin with.

future. This is obviously one of the most risky fields of any science and is often speculative; therefore, I have called this passage "a speculation." But all the same, with today's very detailed models of stellar evolution, it is possible to arrive at relatively precise conclusions about the future of our solar system as well as the stars and galaxies in general. In the last ten to fifteen years we have determined the geometry of the universe and the most important cosmological parameters very well, allowing us to predict the likely course of the cosmic expansion. Due to the accelerated expansion of the cosmos through dark energy, we assume today that the universe will exponentially expand for all eternity. Additionally, the standard model of particle physics and the efforts to combine the general theory of relativity with quantum theory allow us to predict the destiny of matter and of black holes throughout the remarkably long cosmic time span. Because many predictions can only be briefly outlined, I would like to refer to two books that deal with the future of the universe in greater detail: *The Five Ages of the Universe* by Adams and Laughlin[1] and *The Future of the Universe: Chance, Chaos, God?* by Arnold Benz.[2] Figure 9.1 shows a graphical overview of the developmental phases of the universe.

THE FUTURE OF THE SOLAR SYSTEM

The sun formed about 4.55 billion years ago out of a primeval nebula of interstellar matter, which had already been enriched by a number of earlier generations of stars with heavier elements than those produced in the Big Bang. After the sun's interior fusion oven fired up, it continuously produced energy through the nuclear fusion of hydrogen into helium and heavier elements until the hydrogen became depleted. In this way, the chemical elements of the periodic table up as far as the iron atom were "baked" inside the stomachs of the stars.[3] During the phase of hydrogen burning, which is predicted to last roughly eleven billion years, the pressure resulting from the interior heat was balanced by the pressure from the gravitation of the star's mass—the star became stable. But gradually, the sun's chemical composition changed. The core, filled with nuclear "ashes," gradually grew denser and hotter. The radiation pressure intensified, causing the surface of the sun to expand as its luminosity increased. Since its formation, the sun's radiation has already increased noticeably, and by the time it reaches the end of its life, it will have doubled once again.

The increase in insolation has a noticeable effect on Earth and in the end, will ruin the further possibility of life on our planet. Everyone is talking about the danger of the greenhouse effect as a result of global warming. The temperature on Earth has gone up by 0.6°C during the last one hundred years. Experts are still arguing about what percentage is caused by the sun and what part of it is "homemade" and due to human influence. According to recent predictions, scientists assume the temperature will increase by at least 2°C during the next one hundred years. "System Earth" has indeed shown an amazing, and throughout the last millions of years also very successful, ability to compensate for fluctuations in temperature through complex regulation processes of the atmosphere, the biosphere, the hydrosphere, and the geosphere. But in the face of the permanent increase in irradiation caused by the chemical development inside the sun, sooner or later the greenhouse effect on Earth will become unstable. At the Institute for Climate Impact Research at the Astrophysical Institute Potsdam, models of star evolution and detailed climate models were combined to determine the so-called "habitable zone" for Earth and other planets.[4] According to this study, about 500 million to a billion years from now, Earth will grow so hot that all the water will begin to boil, the oceans will evaporate, and the planet will be sterilized. At the same time, however, conditions for the possibility of life on Mars will have improved. A fair amount of time before this happens, it would be advisable for the existing life on Earth to prepare and equip an ark and perhaps populate Mars.

In about seven billion years, when the sun has used up roughly a tenth of its hydrogen supply (and therefore nearly all the hydrogen in its core), the nuclear fusion processes in its interior will accelerate, and the fusion ash will burn into heavier elements. Simultaneously, the core will continue to shrink as the sun's luminosity increases. At this point, it will expand into a red giant, which, due to its size, will be considerably more luminous but considerably cooler than our sun today. Because this star sends out a strong stellar wind, it loses mass and its gravitation is reduced. As a result, the planets will move further out. Earth, for instance, will more or less be located on the momentary orbit of Mars. By then, the sun will have vaporized Mercury and Venus, and, seen from the Earth, its disk will cover a good deal of the firmament. The temperature on Earth will rise to about 1,200°C. By this time, at the latest, Mars will no longer be inhabitable. At the end of the sun's evolution, it will throw off its remaining hydrogen envelope as a so-called planetary nebula—these stellar shells are among the most beautiful and most colorful objects in our Milky Way (see Plate 27). The sun's heavy core will contract and turn into a white dwarf, an object roughly the diameter of Earth and the rest of the sun's mass. A white dwarf is almost entirely composed of "degenerate matter," which is solely stabilized by the degeneracy pressure of the electrons—a consequence of the Heisenberg uncertainty principle and Wolfgang Pauli's exclusion principle.

This star, which will initially shine a bluish white color, contains so much thermal energy that it will radiate for many billions of years while, at the same time, continuously cooling down. Just like our sun, most of the stars in the universe end up as white dwarfs.

THE DESTINY OF THE GALAXIES

Our Milky Way consists of about 200 billion stars—and the interstellar matter, out of which new stars constantly evolve and again disappear—and finally, of the dark matter, which keeps the rapidly rotating stellar system together through gravitational force. The creation of more and more new generations of stars will gradually use up the hydrogen in the galaxies. Inside their "stomachs," the stars fuse hydrogen into heavier elements, which are partly returned to the interstellar medium in the form of stellar winds and supernova explosions and are partly stored in compact leftovers, such as white dwarfs, neutron stars, and stellar black holes.

Our Milky Way is part of a hierarchical structure of small galaxy groups, bigger galaxy clusters, superclusters, and foam-like structures of filaments and voids, around which matter gathered during the formation of the large-scale cosmic structure. Driven by the gravitational attraction of the dark matter, the galaxies merge into bigger and bigger units throughout the development of the cosmos while the universe as a whole continues to expand. The dominant galaxy of our Local Group is the Andromeda Galaxy, toward which the Milky Way is moving at a speed of about 150 kilometers per second. In about four billion years, the Milky Way and the Andromeda Galaxy will near one another and may merge. Such a "galaxy merger" is no threat to the two galaxies. They feel no dramatic changes and their planetary systems remain unimpaired. During this collision, the two black holes in the center of the galaxies feed very efficiently, and as a result, their size dramatically increases. This is how new quasars are born. Just imagine how fantastic and interesting the sky would look in a merging galaxy system. If we wait long enough, all the galaxies in the Local Group and possibly the next big galaxy cluster in the Virgo constellation will merge into one gigantic metagalaxy.

The appearance and life expectancy of stars mainly depends on their mass. Although a star with the mass of our sun, as described above, will shine for about eleven billion years with a temperature of about 5,800 K (in a sort of yellowish-green color), a star of ten solar masses will use up its hydrogen roughly three hundred times quicker and will shine extremely bright with an emission maximum within the ultraviolet spectral range. However, a star with only a tenth of the solar mass will live roughly three hundred times longer than the sun and will emit its light in the red wavelength range. The stars that live the longest are the brown dwarfs, with about 8 percent of the solar mass, which just manage to build up enough pressure in their interior in order to ignite the fusion of deuterium nuclei for a short while. Objects with even less mass turn into big planets. Gas giants, for example, Jupiter, didn't quite succeed at becoming a star.

When the galaxies begin running out of interstellar matter and therefore no new stars can form, the only things left besides the white dwarfs, neutron stars, and black holes will be the red and very old stars. In big elliptical galaxies or within the bulges of spiral galaxies, this situation already applies. In about 10^{13} years, the oldest stars known today will have burned out, and in about 10^{14} years, the normal creation of stars will end. The universe will once again lie in the dark.

THE VERY DISTANT FUTURE

After all this takes place, only the burned-out remains of stars will be left—white dwarfs, neutron stars, and black holes, and also the celestial bodies that were not able to maintain fusion—brown dwarfs and planets. Most of the stellar mass now exists in the form of degenerate matter inside the white dwarfs, which continue to cool over time. The brown dwarfs have been holding on to their remaining hydrogen for a very long time. In extremely rare cases, two brown dwarfs can collide with each other, creating a bigger red dwarf with enough mass to fire up the interior fusion oven. These few normal stars will then be the only source of light in the universe.

If dark matter consists of weakly interacting massive particles, these particles could gradually be captured by the white dwarf's gravitational force, gather inside the stars, and now and again come so close to each other that particles and antiparticles destroy each other. The energy released would stop the white dwarf's cooling process and keep them near the temperature of liquid nitrogen at 65 K for a very long period of time. Following an idea of Freeman Dyson,[5] Adams and Laughlin[6] speculated that due to the extremely long time span and the very high energy supplies within the atmosphere of such white dwarfs, a new biological evolution could take place, perhaps even including the development of complexity and intelligence. The time available for such an evolution is so endlessly long that, despite the relatively low temperatures, considerably more evolutionary steps would be possible than in our civilization. An intelligent species that one day feeds on the energy of dark matter could perhaps evolve much further than we will ever be able to.

The protons will presumably decay on a time scale of about 10^{36} years. As a consequence, all stellar remains will "melt" away, namely, the white and brown dwarfs, neutron stars, and planets. Not, however, the black holes, which keep all forms of matter and energy bound inside their event horizons by their enormous gravitational force. According to Stephen Hawking, stellar black holes will evaporate in about 10^{67} years. Black holes with millions of solar masses, like the one currently in the center of our own galaxy, live roughly 10^{83} years, and black holes with the mass of our whole galaxy live about 10^{98} years. In the course of the almost endlessly available time, the gravitational monster at the center of galaxy M87 will swallow every particle currently located inside the gravitational catchment area of the Virgo Cluster. Such black holes are able to reach hundreds of billions of solar masses and can live about 10^{100} years. The first compact objects that appeared on the stage of the universe will then also be the last to leave.

After roughly 10^{100} years, all the structures will have disappeared out of our universe, which by then will have turned into a state of chaos consisting of radiation and lightweight particles like neutrinos, electrons, and positrons. If the expansion really is exponential up to this point in time, the mean density, which today lies at about one proton/m^3, will by then have been reduced to one positron in the 10^{194}-fold volume of our universe today. The wavelength of the background radiation, which today lies at about one millimeter, will then be about 10^{41} light-years—a truly incredible future. The universe will then again be more or less in the condition out of which it formed. All the energy from the Big Bang will be spread out so thinly across space that it will disappear into nothingness. The universe was formed out of "nothing," and in the end, "nothing" is left. Only, in the meantime, this "nothing" is the highest state of energy that we are aware of today.

Chapter Ten

AND WHAT ABOUT GOD?

After I give a lecture, I am frequently asked: How does God figure in all of this? I used to talk my way out of the question by referring to Pope John Paul II, who, during the rehabilitation of Galileo Galilei in 1992, confirmed that "the error of the theologians of the time . . . was to think that our understanding of the physical world's structure was, in some way, imposed by the literal sense of Sacred Scripture." In his message to the Pontifical Academy of Sciences in 1996, the pope even went a step further and declared: "Some new findings lead us toward the recognition of evolution as more than a hypothesis." And so, finally, we've come to the point at which the Church no longer burns scientists at the stake when their results conflict with the literal interpretation of the Bible. On the contrary: as time moves on, new scientific findings are actually becoming part of the Christian doctrine. But these insights don't seem to reach the faithful as fast as the theologians express them. In January 2003, I gave a lecture in the dialogue series Science—Technology—Religion at the TU Darmstadt on the topic "There Are Far Too Many Stars—Curiosity and Awe in the Field of Astronomy," which a theologian subsequently commented on. I still vividly remember that during his rhetorically excellent, dialectic speech he did not once mention the name of God. In the following discussion he was met with great hostility by part of the audience, who regarded his remarks as too godless. Two years later, after I lectured at the "Deutsches Museum," a man who didn't subscribe to the theory of evolution left the room in protest because he disagreed with the pope's view. Even more alarming, in my opinion, are the subversive activities of the Intelligent Design sects. This might be a good time to mention the excerpt of a letter that Johann Wolfgang von Goethe wrote to his friend, the pastor Johann Caspar Lavater, in 1782:

> You take the gospel as it stands, for the most divine truth: A voice from heaven would not convince me that water burns and that fire quenches, that a woman gives birth without knowing man, and that a dead man rises from the grave. I rather regard this as a blasphemy against the great God and His revelation in Nature.[1]

In mid-2006, the Global Ethic Foundation, inspired by theologian Hans Küng, organized an interesting interdisciplinary symposium titled "Science and Religion" at the University of Tübingen, in the castle of Hohentübingen. Among other discussions, the symposium dealt with Küng's book *The Beginning of All Things*.[2] The aim of the symposium was "to deepen the dialogue between the natural sciences and philosophy/theology on the questions of the origins of the Cosmos, of life and of mankind." Küng asked all the invited scientists and theologians to end each of their presentations with their very personal concept of God, which forced me to look into this matter more thoroughly.

Basically, I am a religious person—hardly surprising, having been born in Oberammergau, the Bavarian village that performs the *Passion Play* every ten years. Back then I even refused military service for religious reasons. However, I base my beliefs mainly on the Sermon on the Mount and the Christian concept of "love thy neighbor." As I prepared for Küng's symposium and read his book, I reflected on my own understanding of the role of God. Küng described the latest discoveries in physics and cosmology with an impressive erudition and detailed knowledge and very meticulously identified the very areas in which our knowledge comes to an end and our understanding grows nebulous—especially regarding the beginning of time and the infinity of space, which I have addressed in the Introduction. In his book, Küng also pointed out that "physical reality is largely inscrutable" and "can ultimately be described only with images, ciphers, comparisons; with models and mathematical formulae."[3] This is where he invites the scientist to go along with the idea of God as a hypothesis, just as theologians go along with intricate and complex scientific findings and models.

To me, it appears somewhat as if Küng is doing what he explicitly criticizes: using God as a stopgap for everything not yet understood. This as a role for God strikes me as very unsatisfying, for it implies that his "territory" is dwindling away day by day. It's true that every new scientific result raises new questions while simultaneously, the knowledge of our ignorance grows bigger, but nevertheless, the boundaries of our knowledge are pushed back more and more. Just a few hundred years ago, God was still allowed to hurl bolts of lightning while today, he would only be responsible for the last unsolved 10^{-35} seconds between the Big Bang and the inflation period. This picture of God bears the danger that, as our understanding increases, he will have to vacate his positions one after the other.

But then what is God? As I read Küng's book, I scribbled down in the margin: "Is God able to move faster than the speed of light? Is he part of the theory of relativity?" Today, we know that our universe must be a great deal bigger than the area accessible to us through electromagnetic signals or gravitational waves within our horizon. So, in order to oversee his "kingdom," God would need much more time than the 13.8 billion years that have passed since the Big Bang. It is indeed nice that science-fiction authors can speculate about wormholes that allow you to tunnel through time and space much faster than the speed of light, but as I have described in Chapter 8, even their inventor, Stephen Hawking, does not take them seriously anymore. Omnipresence all around the universe can therefore only exist beyond causality. But then again, it couldn't have any direct effect on us.

I see God as a deeply spiritual and therefore human phenomenon: without man, no God! On no account do I mean to speak about God in a derogatory way. The general awareness that has developed over millions of years, and the experiences of human history have created a God who plays an extremely important role in the intellectual and emotional self-understanding of civilization. God is inside the human mind—nowhere else. When I expressed this thought, I wasn't yet aware that much more qualified people had already written it down before me—for example, German philosopher Ludwig Feuerbach: "God did not, as the Bible says, make man in His image; on the contrary man, as I have shown in 'The Essence of Christianity,' made God in his image."[4]

The symposium in the castle of Hohentübingen was highly interesting. The prehistorians lectured on the development of the mind, art, music, the belief in the hereafter, and religion in the ancient history of mankind. I was especially fascinated by the sound of a Paleolithic flute discovered in the Geißenklösterle Cave near Blaubeuren by

Tübinger prehistorian Nicholas Conard and his colleagues.[5] The evolutionary biologists, especially the Tübinger Nico Michiels, even went as far as stating that God is part of the Darwinian evolution, mutation, and selection. Civilizations that undergo spiritual strengthening and identification through religious allegiances are more capable of surviving than godless societies. Even Hans Küng basically agreed with these views, though with the reservation that this discussion was only about the "image of God" that mankind itself had created. French philosopher Charles-Lois Baron de Montesquieu said that man created the image of God in his own image: "If triangles had a god, he would have three sides." However, Küng did point out, according to the doctrine of St. Thomas Aquinas, the all-pervasive God who would be there even if nothing else existed—the so-called *ipsum esse,* the "being-in-itself." Here I felt reminded of the mysterious dark energy or the inflaton field at the beginning of the universe; a scalar all-pervading energy field out of which the universe emerged. So perhaps, after all, *Creatio ex nihilo,* the "creation out of nothing," is correct? Are the theologians and the cosmologists not that far apart from each other after all? Is God nothingness?

Appendix

Table 1. Composition of the Universe

Quantity	Formula	WMAP-9 Value	Planck Value
Hubble constant	H_0	69.7 ± 1.4 km/s^{-1} Mpc^{-1}	67.3 ± 1.2 km/s^{-1} Mpc^{-1}
Critical density	$\rho_c = 3\,H_0^2/8\pi\,G$	36.0×10^{-30} g cm^{-3}	33.5×10^{-30} g cm^{-3}
Scalar spectral index	n_s	0.974 ± 0.013	0.9603 ± 0.0073
Total energy density	$\Omega_{tot} = \rho/\rho_c$	1.0023 ± 0.0056	1.0
Dark energy density	Ω_Λ	0.717 ± 0.0028	0.685 ± 0.018
Total matter density	Ω_m	0.283 ± 0.0024	0.315 ± 0.0017
Dark matter density	Ω_{DM}	0.234 ± 0.012	0.265 ± 0.0076
Baryon density	Ω_b	0.0466 ± 0.0018	0.0486 ± 0.0011
Luminous matter	Ω_{stars}	0.0038 ± 0.0020	
Neutrinos	Ω_v	0.0003 to 0.015	
Heavy elements (> He)	Ω_Z	~ 0.0003	

The first eight rows of the table have been taken from the 2014 publication of the Planck data (see P. A. R. Ade et al., "Planck 2013 Results. XV. CMB Power Spectra and Likelihood," *Astronomy & Astrophysics* 571 [2014]: 60), which also compares with the Nine-Year Wilkinson Microwave Anisotropy Probe (WMAP-9) data from Bennet et al., *Astrophysical Journal Supplement Series* 208 (2013): 20. The last rows are taken from the textbook by L. Bergmann and C. Schaefer, *Lehrbuch der Experimentalphysik*, vol. 8, *Sterne und Weltraum* (Berlin: Walter de Gruyter, 2001), 456.

Table 2. The Calendar of the Universe

January 1 00:00	Formation of the elements H, He, . . .
January 1 00:14 a.m.	Matter-radiation decoupling
January 5–13	First stars and black holes are formed
	Stars create the elements C, N, O, . . .
January 20–23	Oldest known galaxies, gamma ray bursts, and quasars
March 27	"Big feast," quasar maximum
September 1	Formation of the sun and Earth
September 4	Oldest minerals on Earth: zircon (Western Australia)
September 14–19	First rock crusts (Canada) and sediment rocks (Greenland)
September 19	First signs of life (chemofossils)
September 29	Oldest fossil remains, bacterial prints (stromatolites)
December 11–15	Snowball Earth
December 16	Exotic, long-extinct organisms (Ediacara fauna)
December 16–19	Big Bang of evolution (Cambrian explosion, all species)
December 20–24	Forest, fishes, reptiles
December 24	Largest mass extinction (70–95 percent of all species die out)
December 25	Mammals develop
December 29	Extinction of the dinosaurs
December 31 8:00 p.m.	Earliest human ancestors *(Australopithecus)*
January 1 12:00 a.m.— 6 minutes	Modern human being *(Homo sapiens)*
–70 seconds	Extinction of the *Neanderthal*
–4.6 seconds	Jesus Christ
–0.23 seconds	Our life span (one hundred years)
January 3	Gravitational waves from NGC 6240
January 12	Earth becomes too hot to live on
April 16	Milky Way is swallowed by the Andromeda Galaxy
July 10	Sun enters the red giant phase

Notes

CHAPTER 1: THE DARK SIDE OF THE UNIVERSE

1. In a contest, Chilean schoolchildren named the four Very Large Telescopes (VLT) at the European Southern Observatory in the Atacama Desert. Using words from the aboriginal Mapuche language, the telescopes were named *Antu,* meaning "sun"; *Kueyen,* meaning "moon"; *Melipal,* meaning "the Southern Cross"; and *Yepun,* meaning "Venus as the Evening Star."

2. Albert Einstein, *The Collected Papers of Albert Einstein,* vol. 8 (Princeton, NJ: Princeton University Press, 1987), Doc. 294, p. 386.

3. Letter from Einstein to Schwarzschild, January 9, 1916 (*Collected Papers of Albert Einstein,* vol. 8, Doc. 181).

4. Jürgen Renn, "The Third Way to General Relativity: Einstein and Mach in Context," in *The Genesis of Relativity,* vol. 3, ed. Jürgen Renn (Boston Studies in the Philosophy of Science, vol. 250) (Dordrecht: Springer, 2007), 945.

5. Julius Scheiner, "Über das Spectrum des Andromedanebels," *Astronomische Nachrichten* 148 (1899): 325.

6. Julius Scheiner, "Note on the Spectrum of the Andromeda Nebula," *Astrophysical Journal* 30 (1909): 69.

7. Hans Oleak, "Scheiners Spektrum des Andromeda Nebels: Über die Natur der Spiralnebel," *Die Sterne* 71 (1995): 95.

8. Carl W. Wirtz, "De Sitters Kosmologie und die Radialbewegungen der Spiralnebel," *Astronomische Nachrichten* 222 (1924): 21.

9. Knut Lundmark, "The Motions and Distances of Spiral Nebulae," *Monthly Notices of the Royal Astronomical Society* 85 (1925): 865.

10. Edwin P. Hubble and Milton Humason, "The Velocity-Distance Relation among Extra-Galactic Nebulae," *Astrophysical Journal* 74 (1931): 43.

11. Steven Weinberg, *The First Three Minutes: A Modern View of the Origin of the Universe* (New York: Basic Books, 1993), 26.

12. John Gribbin, *The Birth of Time: How Astronomers Measured the Age of the Universe* (New Haven, CT: Yale University Press, 2001).

13. Albert Einstein, *Grundzüge der Allgemeinen Relativitätstheorie* (Wiesbaden: Vieweg Verlag, 1982), 126.

14. George Gamow, *My World Line* (New York: Viking Press, 1970).

15. Hendrik B. G. Casimir, "On the Attraction between Two Perfectly Conducting Plates," *Proceedings of the Koninklijke Nederlandse Akademie van Wetenschappen* 51 (1948): 793.

16. F. Chen, Umar Mohideen, et al., "Demonstration of the Lateral Casimir Force," *Physical Review Letters* 88 (2002): 101801–101804.

17. Charles P. Enz and Armin Thellung, "Nullpunktsenergie und Anordnung nicht vertauschbarer Faktoren im Hamiltonoperator," *Helvetica Physica Acta* 33 (1960): 842.

18. Richard Preston, *The First Light: The Search for the Edge of the Universe* (New York: Random House, 1996); Alan Guth, *The Inflationary Universe* (New York: Basic Books, 1998).

19. http://en.wikipedia.org/wiki/Bild, GalacticRotation2_en.svg.

20. Erwin Freundlich, "Über einen Versuch, die von A. Einstein vermutete Ablenkung des Lichtes in Gravitationsfeldern zu prüfen," *Astronomische Nachrichten* 193 (1913): 369.

21. Albrecht Fölsing, *Albert Einstein, Eine Biographie* (Frankfurt am Main: Suhrkamp, 1993), 401.

22. Albert Einstein, "Lens-Like Action of a Star by the Deviation of Light in the Gravitational Field," *Science* 84 (1936): 506. Recently it was discovered in Einstein's private notebooks that he worked on gravitational lenses already in 1912: Jürgen Renn et al., "The Origin of Gravitational Lensing: A Postscript to Einstein's 1936 Science Paper," *Science* 275 (1997): 184.

23. Wolfgang Pauli, *Fünf Arbeiten zum Ausschließungsprinzip und zum Neutrino,* Texte zur Forschung, vol. 27 (Darmstadt: Wissenschaftliche Buchgesellschaft, 1977), 101.

24. Harald Fritzsch, *The Creation of Matter: The Universe from Beginning to End* (New York: Basic Books, 1984).

25. For a while in the early 2000s, the existence of an exotic particle consisting of five quarks—the so-called "pentaquark"—was discussed, but this particle was ruled out by more experiments around 2008. Recently, however, the CERN physicists announced the creation of a very short-lived four-quark particle in their Large Hadron Collider.

26. https://news.brown.edu/articles/2014/02/lux.

27. http://cast.web.cern.ch/CAST/.

Chapter 2: The Big Bang

1. The unit megaparsec (Mpc) is based on the astronomical distance measurement using stellar parallaxes (parallax-second) and corresponds to 3.26 million light-years.

2. The critical density is $\rho_c = 3 \, H_0^2/8\pi G = 10.6 \; 10^{-27}$ kg/m^3 for $H_0 = 75$ km/s/ Mpc.

3. "Two stones," as opposed to "one stone" in the case of Einstein.

4. Alan Guth, *The Inflationary Universe* (New York: Basic Books, 1998).

5. Planck time: 5.39 10^{-44} s, Planck length: 1.61 10^{-35} m, Planck mass: 2.18 10^{-8} kg, Planck temperature: 1.42 10^{32} K.

6. Harald Fritzsch, *The Creation of Matter: The Universe from Beginning to End* (New York: Basic Books, 1984). Fritzsch very nicely describes the history of the particles immediately after the Big Bang.

7. Contrary to the strong anthropic principle, which postulates that our universe was fine-tuned by a Creator exactly in a way that humans can exist.

8. This temperature is about 40 percent lower than the temperature of the microwave background because of the electron-positron-annihilation, which happened after the neutrino-decoupling.

9. Guth, *Inflationary Universe*. Gamow's biography, including the description of his escape from Stalinist Russia, is described in detail.

10. George Gamow, *My World Line* (New York: Viking Press, 1970).

11. Guth, *Inflationary Universe*.

12. R. A. Alpher, H. A. Bethe, and G. Gamow, "The Origin of Chemical Elements," *Physical Review* 73 (1948): 803.

13. Marcus Chown, *The Magic Furnace: The Search for the Origins of Atoms* (New York: Vintage Books, 2000).

14. Gamow, *My World Line*.

15. Steven Weinberg, *The First Three Minutes: A Modern View of the Origin of the Universe* (New York: Basic Books, 1993), 26.

CHAPTER 3: CLEARING UP

1. http://en.wikipedia.org/wiki/Flame.

2. http://en.wikipedia.org/wiki/Olbers%27_paradox.

3. Alan Guth, *The Inflationary Universe* (New York: Basic Books, 1998).

4. A. A. Penzias and R. W. Wilson, "A Measurement of Excess Antenna Temperature at 4080 Mc/s.," *Astrophysical Journal* 142 (1965): 419.

5. R. H. Dicke, P. J. E. Peebles, P. G. Roll, and D. T. Wilkinson, "Cosmic Black-Body Radiation," *Astrophysical Journal* 142 (1965): 414.

6. Rudolf Kippenhahn, *Light from the Depths of Time* (New York: Springer, 1986), 217.

7. A. McKellar, "Molecular Lines from the Lowest States of Diatomic Molecules Composed of Atoms Probably Present in Interstellar Space," *Publications of the Dominion Astrophysical Observatory* 7, no. 15 (1941): 251.

8. J. Delannoy, J. F. Denisse, E. Le Roux, and B. Morlet, "Mesures absolues de faibles densités de flux de rayonnement à 900 MHZ," *Annales d'astrophysique* 20 (1957): 222.

9. G. B. Field and J. L. Hitchcock, "Cosmic Black-Body Radiation at λ=2.6 mm," *Physical Review Letters* 16 (1966): 817.

10. G. F. Smoot et al., "Structure in the COBE Differential Microwave Radiometer First-Year Maps," *Astrophysical Journal Letters* 396 (1992): L1. Most

versions of cosmic inflation predict the scalar index of the Harrison-Zeldovich spectrum to be slightly less than 1, which is consistent with the measured value of 0.96 (see Table 1 in the Appendix).

11. R. A. Sunyaev and Y. B. Zeldovich, "Small-Scale Fluctuations of Relic Radiation," *Astrophysics and Space Science* 7 (1970): 3.

12. P. J. E. Peebles and J. T. Yu, "Primeval Adiabatic Perturbation in an Expanding Universe," *Astrophysical Journal* 162 (1979): 815.

13. Planck collaboration: P. A. R. Ade et al., "Planck 2013 Results. XV. CMB Power Spectra and Likelihood," *Astronomy & Astrophysics* 571 (2014): 60.

14. Josef Silk, *The Infinite Cosmos: Questions from the Frontiers of Cosmology* (New York: Oxford University Press, 2008).

15. This mission combined two competing projects: The Cosmic Background Radiation Anisotropy Satellite (COBRAS) was proposed by Nazzareno Mandolesi from Bologna University. One of the team members, George Smoot, was principal investigator of the COBE DMR experiment. French astronomer Jean-Loup Puget led the other project, called Satellite for Measurements of Background Anisotropies (SAMBA).

16. C. L. Bennett et al., "Microwave Anisotropy Probe: A MIDEX Mission Proposal," 1996, http://map.gsfc.nasa.gov/.

17. E. J. Wollack, N. Jarosik, C. B. Netterfield, L. A. Page, and D. Wilkinson, "A Measurement of the Anisotropy in the Cosmic Microwave Background Radiation at Degree Angular Scales," *Astrophysical Journal Letters* 419 (1993): L49.

18. A. D. Miller, R. Caldwell, M. J. Devlin, W. B. Dorwart, T. Herbig, M. R. Nolta, L. A. Page, J. Puchalla, E. Torbet, and H. T. Tran, "A Measurement of the Angular Power Spectrum of the Cosmic Microwave Background from L=100 to 400," *Astrophysical Journal* 524 (1999): L1.

19. P. de Bernardis et al., "A Flat Universe from High-Resolution Maps of the Cosmic Microwave Background Radiation," *Nature* 404 (April 27, 2000): 955.

CHAPTER 4: THE COSMIC WEB OF GALAXIES

1. John P. Huchra, "Mapmaker, Mapmaker Make Me a Map," in *Our Universe*, ed. S. Alan Stern (New York: Cambridge University Press, 2001), 7.

2. M. J. Geller and J. P. Huchra, "Mapping the Universe," *Science* 260 (1989): 1175.

3. H. Ebeling, C. R. Mullis, and R. B. Tully, "A Systematic X-Ray Search for Clusters of Galaxies behind the Milky Way," *Astrophysical Journal* 580 (2002): 774.

4. R. B. Tully, Y. Courtois, D. Hoffman, and D. Pomarède, "The Laniakea Supercluster of Galaxies," *Nature* 513 (2014): 71.

5. "Hawaii Scientist Maps, Names Laniakea, Our Home Supercluster of Galaxies," Institute for Astronomy, University of Hawai`i, September 3, 2014, http://www.ifa.hawaii.edu/info/press-releases/Laniakea/.

6. 2 Micron All Sky Survey, performed by the Infrared Processing and Analysis Center, California Institute of Technology and the University of Massachusetts.

7. http://pan-starrs.ifa.hawaii.edu/public/.

8. http://www.dlr.de/pf/desktopdefault.aspx/tabid-8951/15456_read-37907/, October 14, 2013.

9. http://www.dlr.de/desktopdefault.aspx/tabid-830/1331_read-2408/.

10. http://en.wikipedia.org/wiki/Tunguska_event.

11. http://www.astronews.com/news/artikel/2003/09/0309–011.shtml.

12. Matt. 25:29.

13. A. Kravtsov and A. Klypin (National Center for Supercomputer Applications, University of Chicago), S. Gottlöber (Leibniz Institute for Astrophysics, Potsdam). The corresponding movies can be found under http://cosmicweb.uchicago.edu/images/mov/.

14. V. Springel, C. S. Frenk, and S. D. M. White, "The Large-Scale Structure of the Universe," *Nature* 440 (2006): 1137. Movies can be downloaded from the website: http://www.mpa-garching.mpg.de/galform/virgo/millennium/.

15. In the meantime, Volker Springel moved on to a professorship at the Institute for Theoretical Astrophysics at the University of Heidelberg.

16. V. Springel, "The Cosmological Simulation Code GADGET-2," *Monthly Notices of the Royal Astronomical Society* 364 (2005): 1105, http://www.mpa-garching.mpg.de/galform/virgo/millennium/. Movies can be downloaded.

17. Unfortunately, at the beginning of 2007 the Advanced Camera for Surveys (ACS) failed due to a problem in the power supply. In a heroic effort and after many years of delay—not least due to the loss of the space shuttle *Columbia* in 2004—NASA flew the final and most challenging Hubble Servicing Mission 4 (SM4) on the space shuttle *Atlantis* in May 2009. The astronauts on this mission not only repaired the ACS but also installed two new powerful instruments, the Wide-Field Camera 3 (WFC3) and the Cosmic Origins Spectrograph (COS).

18. R. Ekers, Commonwealth Scientific and Industrial Research Organization (CSIRO); C. Cesarsky, European Southern Observatory (ESO); G. Hasinger, Max Planck Institute for Extraterrestrial Physics (MPE); G. Illingworth, University of California, Santa Cruz (NCSC); J. Mould, National Optical Astronomy Observatory (NOAO); M. Mountain, Gemini Observatory; A. Sargent, California Institute of Technology (Caltech); T. Soifer, Caltech; H. Tannenbaum, Harvard-Smithsonian Center for Astrophysics (CfA); B. Williams, Space Telescope Science Institute (STScI); and R. Windhorst, Arizona State University, http://www.stsci.edu/hst/udf/adv_cmte.

19. ESO image: http://www.eso.org/outreach/press-rel/pr-2000/phot-07a-00-normal.jpg; Spitzer image: http://www.spitzer.caltech.edu/images/1419-ssc2005-11a-Spitzer-Spies-Spectacular-Sombrero.

20. http://www.weltderphysik.de/de/1120.php.

21. M. Steinmetz, Leibniz Institute for Astrophysics, Potsdam, http://www
.aip.de/Members/msteinmetz/movies.

Chapter 5: Star Formation

1. Tom Abel, Greg L. Bryan, and Michael L. Norman, "The Formation of
the First Star in the Universe," *Science* 295 (2002): 93; Volker Bromm and Richard B Larson, "The First Stars," *Annual Review in Astronomy and Astrophysics*
42 (2004): 79.

2. T. A. Rector (University of Alaska, NRAO/AUI/NSF and NOAO/AURA/
NSF) and B. A. Wolpa (NOAO/AURA/NSF), The Hubble Heritage Project, http://
heritage.stsci.edu/2005/12b/supplemental.html.

3. NASA, ESA, and The Hubble Heritage Team (STScI/AURA), http://
heritage.stsci.edu/2005/12b/big.html.

4. João F. Alves (European Southern Observatory), Charles J. Lada (Harvard-
Smithsonian Center for Astrophysics), and Elizabeth A. Lada (University of Florida), http://www.eso.org/outreach/press-rel/pr-2001/pr-01-01.html.

5. "ALMA Takes Close Look at Drama of Starbirth," Phys.org, August 20,
2013, http://phys.org/news/2013-08-alma-drama-starbirth.html.

6. Mark McCaughrean (Max Planck Institute for Astronomy), C. Robert
O'Dell (Rice University), and NASA, "Panoramic Hubble Picture Surveys Star
Birth, Proto-Planetary Systems in the Great Orion Nebula," http://hubblesite.org
/newscenter/newsdesk/archive/releases/1995/45/image/b.

7. J. Bally (University of Colorado), H. Throop (Southwest Research Institute, Boulder), C. R. O'Dell (Vanderbilt University), and NASA/European Space
Agency (ESA), "Protoplanetary Disks in the Orion Nebula," http://www
.spacetelescope.org/images/html/opo0113a.html.

8. NASA, ESA, M. Livio, and the Hubble Twentieth Anniversary Team (Space
Telescope Science Institute), "Starry-Eyed Hubble Celebrates 20 Years of Awe and
Discovery," http://www.nasa.gov/mission_pages/hubble/science/hubble20th-img
.html.

Chapter 6: Wanderers in the Sky

1. George Biddell Airy, "Letter of Mr. Airy, Astronomer Royal, to the Editor
(Schreiben des Herrn Airy, Astronomer Royal, an den Herausgeber)," *Astronomische Nachrichten* 15 (1838): 217.

2. George Biddell Airy, "Account of Some Circumstances Historically Connected with the Discovery of the Planet Exterior to Uranus," *Astronomische Nachrichten* 25 (1847): 133.

3. Ibid., 149.

4. Nick Kollerstrom, "Neptune's Discovery, The British Case for Co-
Prediction," University College London, 2001, http://www.ucl.ac.uk/sts/nk/neptune.

5. This refers to the so-called ecliptic coordinates, along and perpendicular
to the plane of the solar system.

6. Galle to Le Verrier, letter, Berlin, September 25, 1846, in *Michigan Journal of Education* 6 (1859): 370.

7. Kollerstrom, "Neptune's Discovery."

8. The asteroid belt beyond Neptune was actually discovered at the eighty-eight-inch telescope on Maunakea, Hawaiʻi, and was named in honor of Gerard Kuiper, who discovered the excellent astronomical utility of this mountain in 1964. David Jewitt and Jane Luu, "Discovery of the Candidate Kuiper Belt Object 1992 QB_1," *Nature* 362 (1993): 730–732.

9. F. Bertoldi, W. Altenhoff, A. Weiss, K. M. Menten, and C. Thum, "The Trans-Neptunian Object UB_{313} Is Larger than Pluto," *Nature* 439 (2006): 563.

10. The mendicant monk Giordano Bruno was a free spirit whose beliefs in many ways disagreed with the doctrine of the Catholic Church. Until the end, he denied the divine sonship of Jesus Christ and the existence of the Final Judgment and held to his conviction regarding the existence of a multitude of alien worlds. He was expelled from the Church as well as from the Dominican Order and sentenced to death at the stake for heresy and magic. Prior to his execution, his tongue was allegedly bound so he could not speak to those present. In 2000, the Catholic Church designated his execution as a "wrongdoing."

11. The 2015 statistics of exoplanets discovered through different methods include: direct imaging, 53; microlensing, 34; radial velocity or astrometry, 599; transit, 1,199; pulsar, 18; transit timing variability, 3. http://exoplanet.eu/catalog.php.

12. L. A. Buchhave, D. W. Latham, A. Johansen, et al., "An Abundance of Small Exoplanets around Stars with a Wide Range of Metallicities," *Nature* 486 (2012): 375–377.

13. "UH Astronomer Finds Planet in the Process of Forming," Institute for Astronomy, University of Hawaii, October 19, 2011, http://www.ifa.hawaii.edu/info/press-releases/formingplanet/.

14. "The First Double-Double: Astronomers Find Two Planets Orbiting a Two-Star System," Institute for Astronomy, University of Hawaii, August 28, 2012, http://www.ifa.hawaii.edu/info/press-releases/Kepler47/.

15. "A Strange Lonely Planet Found without a Star," Institute for Astronomy, University of Hawaii, October 9, 2013, http://www.ifa.hawaii.edu/info/press-releases/LonelyPlanet/.

16. "Scientists Find Earth-Sized Rocky Exoplanet," Institute for Astronomy, University of Hawaii, October 30, 2013, http://www.ifa.hawaii.edu/info/press-releases/Kepler-78b/.

17. "Astronomers Conclude Habitable Planets Are Common," Institute for Astronomy, University of Hawaii, November 4, 2013, http://www.ifa.hawaii.edu/info/press-releases/HabitablePlanetsCommon/.

Chapter 7: The Stellar Cemetery

1. Rudolf Kippenhahn, *100 Billion Suns: The Birth, Life and Death of the Stars* (New York: Basic Books, 1983).

2. The phase space is the quantum mechanics totality of all possible spatial coordinates, velocity vectors, and spin states that a particle can assume.

3. W. Baade and F. Zwicky, "Cosmic Rays from Supernovae," *Proceedings of the National Academy of Sciences of the United States of America* 20 (1934): 259.

4. J. R. Oppenheimer and G. M. Volkoff, "On Massive Neutron Cores," *Physical Review* 55 (1939): 374.

5. F. S. Kitaura, H.-Th. Janka, and W. Hillebrandt, "Explosions of O-Ne-Mg Cores, the Crab Supernova, and Subluminous Type II-P Supernovae," *Astronomy & Astrophysics* 450 (2006): 345, http://www.mpa-garching.mpg.de/mpa/research /current_research/hl2006-7/hl2006-7-de.html.

6. R. Sunyaev et al., "Discovery of Hard X-ray Emission from Supernova 1987A," *Nature* 330 (1987): 227.

7. http://en.wikipedia.org/wiki/Crab_Nebula.

8. The Crab Nebula in Taurus, November 17, 1999, http://www.eso.org /public/images/eso9948f/.

9. "Space Movie Reveals Shocking Secrets of the Crab Pulsar," September 19, 2002, http://chandra.harvard.edu/photo/2002/0052/0052_xray_opt.jpg.

10. B. Aschenbach and W. Brinkmann, "A Model of the X-ray Structure of the Crab Nebula," *Astronomy & Astrophysics* 41 (1975): 147.

11. J. Trümper, G. Hasinger, B. Aschenbach, H. Bräuninger, U. G. Briel, W. Burkert, H. Fink, E. Pfeffermann, W. Pietsch, P. Predehl, J. H. M. M. Schmitt, W. Voges, U. Zimmermann, and K. Beuermann, "X-ray Survey of the Large Magellanic Cloud by *ROSAT*," *Nature* 349 (1991): 579.

12. The actual observations show that the total amount of energy varies between different explosions. But there is a tight correlation between the duration of the explosion and its total energy so that these differences can be calibrated out. The type Ia explosions are therefore "standardizable."

13. Neutron stars, like terrestrial planets, have a solid crust floating on a liquid core. Similar to the earthquakes originating from plate tectonics on Earth, the neutron star crust can undergo starquakes, which are much more energetic. These processes can indeed lead to explosive emissions of gamma and X-ray radiation. This phenomenon, however, only explains a small fraction of gamma ray bursts, the "soft gamma repeaters."

14. E. Costa, F. Frontera, J. Heise, et al., "Discovery of an X-ray Afterglow Associated with the Gamma-Ray Burst of 28 February 1997," *Nature* 387 (1997): 783.

15. J. van Paradijs, P. J. Groot, T. Galama, et al., "Transient Optical Emission from the Error Box of the Gamma-Ray Burst of 28 February 1997," *Nature* 386 (1997): 686.

16. My colleague Jochen Greiner maintains a very useful and informative website on gamma ray bursts: http://www.mpe.mpg.de/~jcg/grbgen.html.

CHAPTER 8: COSMIC MONSTERS

1. The escape velocity can be easily calculated from the balance between the kinetic energy of a body with mass m and velocity v: $(E_{kin} = 1/2\ m\ v^2)$ and its potential energy in the gravity field of a celestial body with mass M: $(E_{pot} = GMm/r)$. It depends on the square root of the ratio M/r.

2. The exact value of the velocity of light is 299,792,458 meters per second.

3. Kip S. Thorne, *Black Holes and Time Warps* (New York: W. W. Norton, 1995).

4. S. D. Mathur, "The Quantum Structure of Black Holes," *Classical and Quantum Gravity* 23 (2006): 115.

5. S. Gillessen et al., "Fifteen Years of High Precision Astrometry in the Galactic Center," *IAU Symposium* 248 (2008): 466.

6. R. Genzel, F. Eisenhauer, and S. Gillessen, "The Galactic Center Massive Black Hole and Nuclear Star Cluster," *Reviews of Modern Physics* 82 (2010): 3121.

7. S. Gillessen et al., "A Gas Cloud on Its Way towards the Supermassive Black Hole at the Galactic Centre," *Nature* 481 (2012): 51; S. Gillessen et al., "Pericenter Passage of the Gas Cloud G2 in the Galactic Center," *Astrophysical Journal* 774 (2013): 44.

8. The maser effect is similar to the related laser effect, which strongly amplifies and efficiently focuses visible light.

9. K. Gültekin et al., "The M–σ and M–L Relations in Galactic Bulges, and Determinations of Their Intrinsic Scatter," *Astrophysical Journal* 698 (2009): 198.

10. S. Gezari, "The Tidal Disruption of Stars by Supermassive Black Holes," *Physics Today* 67 (2014): 37.

11. http://nobelprize.org/nobel_prizes/physics/laureates/2002/giacconi-au tobio.html.

12. R. Giacconi, G. W. Clark, and B. B. Rossi, "A Brief Review of Experimental and Theoretical Progress in X-ray Astronomy," *Technical Note of American Science & Engineering*, ASE-TN-49 (January 15, 1960).

13. Riccardo Giacconi, Herbert Gursky, Frank R. Paolini, and Bruno B. Rossi, "Evidence for X-rays from Sources Outside the Solar System," *Physical Review Letters* 9 (1962): 439.

14. G. Hasinger and M. van der Klis, "Two Patterns of Correlated X-ray Timing and Spectral Behaviour in Low-Mass X-ray Binaries," *Astronomy & Astrophysics* 225 (1989): 79.

15. Riccardo Giacconi and Bruno Rossi, "A 'Telescope' for Soft X-ray Astronomy," *Journal of Geophysical Research* 65 (1960): 773.

16. J. Trümper, W. Pietsch, C. Reppin, W. Voges, R. Staubert, and E. Kendziorra, "Evidence for Strong Cyclotron Line Emission in the Hard X-ray Spectrum of Hercules X-1," *Astrophysical Journal* 219 (1978): L105.

17. Werner Heisenberg, "Schritte über Grenzen," in *Gesammelte Reden und Aufsätze* (München: R. Piper, 1971).

18. Hoimar von Ditfurth, *Im Anfang war der Wasserstoff* (Hamburg: Hoffmann & Campe, 1972).

19. Rudolf Kippenhahn, *100 Billion Suns: The Birth, Life and Death of the Stars* (New York: Basic Books, 1983).

20. J. H. M. M. Schmitt, S. L. Snowden, B. Aschenbach, G. Hasinger, E. Pfeffermann, P. Predehl, and J. Trümper, "A Soft X-ray Image of the Moon," *Nature* 349 (1991): 583.

21. Some publications about the resolution of the X-ray background into discrete sources with ROSAT, XMM-Newton, and Chandra: G. Hasinger, R. Burg, R. Giacconi, G. Hartner, M. Schmidt, J. Trümper, and G. Zamorani, "A Deep X-ray Survey in the Lockman Hole and the Soft X-ray Log N—Log S," *Astronomy & Astrophysics* 275 (1993): 1; G. Hasinger, R. Burg, R. Giacconi, M. Schmidt, J. Trümper, and G. Zamorani, "The ROSAT Deep Survey I. X-ray Sources in the Lockman Field," *Astronomy & Astrophysics* 329 (1998): 482; G. Hasinger, B. Altieri, M. Arnaud, X. Barcons, J. Bergeron, et al., "XMM-Newton Observation of the Lockman Hole. I. The X-ray Data," *Astronomy & Astrophysics* 365 (2001): L45; R. Giacconi, P. Rosati, P. Tozzi, M. Nonino, G. Hasinger, et al., "First Results from the X-ray and Optical Survey of the Chandra Deep Field South," *Astrophysical Journal* 551 (2001): 624; W. N. Brandt, A. E. Hornschemeier, D. M. Alexander, G. P. Garmire, D. P. Schneider, et al., "The Chandra Deep Survey of the Hubble Deep Field North Area. IV: An Ultradeep Image of the HDF-N," *Astronomical Journal* 122 (2001): 1.

22. Richard Preston, *First Light: The Search for the Edge of the Universe* (New York: Random House, 1996).

CHAPTER 9: THE DESTINY OF THE UNIVERSE

1. Fred Adams and Greg Laughlin, *The Five Ages of the Universe: Inside the Physics of Eternity* (New York: Free Press, 2000).

2. Arnold Benz, *The Future of the Universe: Chance, Chaos, God?* (London: Continuum, 2002).

3. Elements heavier than iron are formed only in supernova explosions, the violent deaths of massive stars.

4. S. Franck, W. von Bloh, C. Bounama, M. Steffen, D. Schönberner, and H.-J. Schellnhuber, "Determination of Habitable Zones in Extrasolar Planetary Systems: Where Are Gaia's Sisters?," *Journal of Geophysical Research* 105 (2001): 1651.

5. F. J. Dyson, "Time without End: Physics and Biology in an Open Universe," *Reviews of Modern Physics* 51 (1979): 447.

6. Ibid.

CHAPTER 10: AND WHAT ABOUT GOD?

1. Letter from Johann Wolfgang Goethe to Johann Caspar Lavater, August 9, 1782, cited in Friedrich Bruns, "Goethes 'Grenzen der Menscheit,'" *Journal of English and Germanic Philology* (University of Illinois, Urbana, Illinois) 18 (1919): 100.

2. Hans Küng, *The Beginning of All Things: Science and Religion* (Grand Rapids, MI: Wm. B. Eerdmans, 2008), http://www.weltethos.org/00-symposion .htm.

3. I am of the opinion that this statement applies to all realities (see the Introduction).

4. Ludwig Feuerbach, *Das Wesen der Religion* (Leipzig, Germany, 1846).

5. http://www.sueddeutsche.de/imperia/md/audio/wissen/shaman.mp3.

Index

Production Notes for Hasinger / *Astronomy's Limitless Journey*

Cover design by Julie Matsuo-Chun

Composition by Westchester Publishing Services with
display type and text in Sabon Hawn.

Printing and binding by Sheridan Books, Inc.

Printed on 55 lb. House White Hi-Bulk D37, 360 ppi.